THE ROCKHOUNDING BIBLE

5 in 1

THE MOST COMPLETE GUIDE TO FINDING, IDENTIFYING, AND COLLECTING PRECIOUS GEMS, MINERALS, GEODES, AND FOSSILS

TYLER GORDON

TABLE OF CONTENTS

TABLE OF CONTENTS

INTRODUCTION

This book will teach you how to identify gemstones, minerals, geodes, and fossils, as well as their current and historical uses in many civilizations and societies. They can be used for various purposes, including enhancing spirituality and meditation, collecting them as a pastime, or simply for fashion. It's entirely up to you how you use them.

If you don't broaden your understanding of these delightful gems, you risk misidentifying them or misusing them. Before venturing into the world of precious stones, it is critical to arm yourself with as much knowledge as possible. Aside from being functional, these stones are steeped in mythology and folklore. Whether you believe in their mystical properties, they can provide insight into different cultures and worldviews.

This guide is for you if you want a good time but are overwhelmed by the information available. I'll tell you everything you need to know about getting ready for your first Rockhound adventure. It includes basic research, tools, and important regulations you should know.

WHAT IS ROCKHOUNDING?

The pursuit of minerals, stones, crystals, and gems is known as rockhounding. Rockhounding is a popular pastime that appeals to people of all ages. The fact that hobbyists can search different natural sites for the specimen they want contributes to the swing's popularity.

Rockhounding appeals to people for a variety of reasons. Jewelers may be looking for stones to use in their designs, and people interested in local history and geology may be interested in getting hands-on experience with the minerals available. Some could even desire to add difficulty to their adventure. Whatever the motive for partaking in rockhounding, many people find it enjoyable and useful.

Doing some research is the first step in starting a rock-hounding pastime. Here's what you need to know:

- In your area, where are the most popular rock hunting spots?
- What minerals can you find in this area?
- Is there a local rock and mineral club?

You're ready to start organizing the logistics of your first rockhounding journey once you've sorted out the answers to these questions. Find out where you may find rocks in your area and where you can go rockhounding.

Like many other things, the internet may be your first source of information about the best Rockhound sites in your area. While it may be tempting to review websites that provide vast information from all over the United States or the world, narrowing your search to a specific location will yield more accurate results.

Stones can be found on public or private land and in professional excavations. There are several public states where people can gather rocks as a hobby, but there are some that are not. Access to private property is only permitted with the owner's permission. Paid excavators and miners understand what's on their property and may steer beginners in the right way while hunting for interesting rocks. I propose looking at the official stones, minerals, and jewels in your state and adjacent countries, as these are easier to find since these documents are often official (though not always).

There are also top export and local names for the same natural resource. While the internet can help you get started, I also recommend conducting extensive research using the methods listed below:

Books and Magazines

Public libraries should also be on your list of places to visit if you want to learn more about your new pastime. The library may have information relevant to the geology of your city or nation in books or journals.

Librarians are priceless assets. So, if you can't locate what you're looking for, get in touch with the help desk. They are eager to assist you and point you toward resources that are not immediately apparent. Local bookstores are more likely than large businesses to carry region-specific titles. When books are out of stock, they can always assist you in locating and ordering them.

Some bookstores specialize in rare or unpublished books, giving them a great place to look for hard-to-find items. Rock & Gem Magazine is the most widely circulated national publication for rock greyhounds in the United States. They publish a calendar of events and a list of clubs on their website and different articles on the subject. Other periodicals dealing with natural history or related themes may also have articles on this subject.

Join a Rock and Mineral Club in Your Area

Joining a rock club is a great way to learn about rockhounding in your area while meeting new people who share your interests. They are also an excellent place to learn about the equipment you will need from people who have used it. Other members can also rent or sell you equipment.

Start with the American Federation of Mineralogy Societies if you're looking for a rock club. Their website includes a list of local groups and links to regional websites. Most club lists will contain regular meeting hours and locations, but we recommend confirming with an authorized representative.

Here are some additional resources for learning about rocking:

NATURAL RESOURCES AGENCY OF THE STATE

Agencies in almost every state manage geology and mining. Depending on where they are located, they may have different names, but they usually have a lot of information about local gems, minerals, and fossils.

Although they may not have a public-facing website, they may usually supply information upon request, and some have a robust online presence.

VISIT A MUSEUM

Around the country, various geological and science museums can be found. They not only show you the possible specimens in the area, but they also have a regular training program.

UNIVERSITY IN THE AREA

You can locate resources to assist your interest in rock activity if you live nearby or attend a university with a related curriculum (geology, mineral engineering, paleontology).

SHOP FOR STONES

Visiting a store specializing in stones, gems, and minerals is a terrific way to meet new people and examine high-quality samples of items you're interested in.

Local rock hunting areas and rock clubs are frequently known to the staff. A modest number of local rock samples may be sold in a tourist-oriented store, but you're more likely to find a lesser-known vendor than in a specialty store.

There is usually a small entrance fee for these one-time events, and you can find a wide variety of exceptional specimens. Even though the vendors are the main draw, the show always includes demos, exhibitions, and performances.

 If you have children, bring them rockhounding as soon as possible. Most young children enjoy playing with rocks, so they'll find it fun!

What's the difference between a rock, a mineral, and a gem? These three phrases are frequently used interchangeably in rockhounding; however, there are some distinctions:

- **Rock**: A rock is a mineral aggregate (composition of minerals) that does not require a particular chemical makeup. Sedimentary, igneous, and metamorphic rocks are the three types of rock.

- **Mineral**: A mineral is a solid, inorganic, crystalline material that occurs naturally and has a specific chemical makeup. Minerals appear in thousands of forms, whereas rocks come in three basic categories.

- **Gem**: Any rock or mineral with beauty and economic worth is referred to as a gemstone.

The ancient Egyptians used minerals and their inventive cosmetics in various ornaments and decorations. King Tutankhamen's tomb, for example, was discovered to be brimming with not only gold but also numerous gemstone treasures. The young king's solid gold burial mask was inlaid with lapis lazuli, obsidian, quartz, carnelian, amazonite, and turquoise. A scarab carving made of Libyan desert glass, a type of tektite, was also discovered on a breastplate in the tomb.

A BRIEF HISTORY OF ROCKHOUNDING

Since the beginning of time, people have collected and treasured rocks and minerals and the various jewels and fossils found within them. Who wouldn't be enthralled by the earth's natural wonders, from sparkling crystals to valuable stones to fossilized animal footprints?

Around the world, archaeologists have uncovered mines where ancient civilizations harvested various rocks and minerals. To date, the oldest known mine is thought to be roughly 43 thousand years old. It was a hematite deposit used to create ocher, a red pigment used in paint and cosmetics. Lapis lazuli and turquoise are currently mined in Afghanistan and Iraq, respectively, and are estimated to be 5,000 years old. And as early as 4300 B.C., Neanderthal sites in Belgium revealed flint mining indications.

Early Applications of Rocks and Minerals

What precisely did these ancient people do with their looted treasures? While rocks and minerals have been used in various ways throughout history, the most typical early uses include manufacturing tools, weapons, cosmetics, jewelry, other aesthetics, and medicine. The sections that follow go over each of these applications in further depth.

Tools

The oldest known tools were discovered in Kenya around 3.3 million years ago. These knapped rock tools are more than 700,000 years older than the stone tools previously thought to be the oldest in human history, which were discovered in Ethiopia. What these antique tools were utilized for is unknown.

The history and extent of man-made tools and the various rocks and minerals they were formed from are immense. Early civilizations around the world were inventive with the hidden prizes native to their land, as evidenced by artifacts and records left behind. Flint was widely mined throughout Europe at the end of the Stone Age (twelve thousand years ago) for its durability in many implements, such as axes, and its ability to start a flame. Vikings in Scandinavia (from the eighth to eleventh centuries) are thought to have built an early version of an optical lens out of quartz crystal and employed Iceland spar, a type

of optical calcite, as a navigational tool on their boats. Many cultures have long used obsidian, an incredibly sharp mineral (technically a mineraloid, a mineral-like material), to manufacture tools such as knives and scraping instruments for cleaning animal hides. It was also originally the predominant material in eye surgery scalpels. Obsidian scalpels are still used by some surgeons nowadays.

Weapons

The practical applications of rocks and minerals were not limited to tools. They were turned into weapons almost as soon as they were devised for non-lethal purposes like cutting and carving wood. Since the early Stone Age, sharp obsidian has been used as blades, arrowheads, and more. Weapons made of flint were also common.

In 2016, researchers determined that a dagger recovered in the tomb of Egyptian King Tutankhamen in 1925 was made of meteorite or space material pieces.

Jewelry, Cosmetics, and other Accessories

It didn't take long for people to understand that rocks and minerals could be used for more than just making tools and weapons. Since the dawn of time, evidence suggests that recovered specimens have been used for aesthetic reasons such as jewelry, cosmetics, and more.

Bead drilling dated back one million years and was one of the first artisanal products made from rocks and minerals. Early Hindu and Buddhist rosaries were made from minerals such as agate and cinnabar. Traditional Japanese beads, known as magatamas, were initially made of slate and talc before becoming almost entirely of jade. Garnet was one of the first faceted beads, with Egyptian bead makers cutting, polishing, and piercing the stone by 3100 B.C. Bracelets and necklaces were typically made by stringing beads together (or independently as early pendants). According to archeological evidence, small fossils were also worn as beautiful pendants by Stone Age people.

Ancient innovators also discovered ways to use minerals to beautify their skin. Metal sulfides like galena, pyrolusite, magnetite, and stibnite were used to create the famed Egyptian black eye cosmetics kohl. Many of these minerals are poisonous to humans; thus, "kohl" is now manufactured with synthetic components that are safe to wear—ancient Egyptians and Sumerians employed copper minerals such as malachite and chrysocolla to manufacture green cosmetics.

Medicine

Minerals were also used medicinally by the ancient Egyptians. Malachite and chrysocolla were great for treating various diseases, such as stomach and dental problems. These minerals were also used to make copper powder, which was used to clean wounds. Since then, copper has been shown to have antibacterial properties.

Many people also used gems and minerals as talismans to ward off physical and spiritual diseases. The early Chinese wore polished jade amulets to ensure health and longevity. In the Baltic region, amber amulets were thought to protect against diseases, and powdered amber was used to aid wound and infection healing. In medieval Western Europe, sapphire amulets were worn, or powdered sapphire was consumed to relieve the pain of rheumatism. In early China and medieval Europe, taking powdered fossils was thought to extend one's life.

Medicine

Fossils offer a glimpse into the past that was lost by early civilizations. They are more than just decorations or the promise of a longer life. According to some researchers, the Greek mythological monsters were inspired by the fossil remains of massive prehistoric creatures. Interest in the origins of fossils sparked the birth of paleontology, the study of life-forms from previous geologic periods in the early 1800s. The preserved footprint of a once-living species was as fascinating to these early rockhounds, collectors, and accidental discoverers as a brilliant stone.

Where do you look for various rocks, minerals, and mineraloids? When rockhounding, what should you bring with you? What methods do you use to clean and store your finds?

There's much more to rockhounding than meets the eye, from locating a specific fossil or gemstone to preparing it for display or wearing it as jewels. You may be wondering where to start. Rockhounding is more than just a fun pastime. It's a time-honored custom that dates back to the dawn of time! The world's first civilizations sought out rocks and minerals to use in everything from weaponry to cosmetics.

In this chapter, you'll discover more about rockhounding's illustrious history and how that history has influenced present tactics for locating and exploiting the planet's hidden gems. You'll also learn why rockhounding is such a popular activity today and what may have prompted you to learn more.

Rockhounding is a hobby that attracts people for as many reasons as there are rocks and minerals. Rockhounding is an outdoor hobby that provides good (and occasionally dirty) entertainment. It is likely to be right up your alley if you appreciate being outside in nature.

Some people plan their entire trip around rockhounding at one or more locations. It is occasionally used as a diversion or supplement to traditional outdoor activities such as fishing or hunting due to their proclivity for wandering off the beaten path when tracking wildlife. Others may spend their free time looking for rocks while on vacation or business.

Other outdoor hobbies can also be performed as a result of rockhounding. Some rockhounds will organize their searches near hot springs, waterfalls, or other natural beauties to unwind after searching for specimens with a peaceful soak, refreshing swim, or different enjoyable experience.

Rockhounding is a terrific lesson in patience and having fun whether or not you locate the rocks and minerals you're looking for. The patience comes from prospecting and gathering for a long time and the reality that you may put in a lot of labor for little reward.

You may think you've been skunked when you don't find any specimens. Accepting that this will occasionally happen to you is one of the most fundamental aspects of rockhounding. As a result, it's critical to value both the process and the limited opportunities rockhounding provides. The natural scenery and fresh air are enough to make an experience enjoyable. Don't give up if you're not finding the same quality or quantity of content as others in clubs or online groups. Like many other things, you'll only get better with practice. Try not to compare your adventures to those of other rockhounds.

What exactly are you seeking out there? On Earth, dozens of rock kinds and over 5,000 minerals have been discovered, with many having distinct variants. Thousands of fossils have also been uncovered, with new species being found regularly. In a nutshell, you can collect various rocks, minerals, and fossils!

It's a good idea to spend some time finding a variety of rocks and minerals when you first start rockhounding. You might find that you enjoy looking for them all. You might discover that agates fascinate you the most or that finding fossils is your calling. Try a little bit of everything to see what interests you the most.

You can start collecting specific items once you've figured out what you like. You may devote all your time to collecting a single mineral in all its crystal forms or a single mineral from several sources. Some people enjoy collecting huge specimens, while others enjoy collecting micro amounts, also known as microscopic specimens. Some people are unconcerned about the type of mineral they find as long as it is heart-shaped. Many people collect stones exclusively for lapidary purposes, such as producing cabochons, spheres, or faceting. The varieties of rock and mineral collections you can make are endless.

Mine Dump Collecting

This is the most straightforward and most efficient method of collecting. As previously noted, the minerals and stones we seek today are frequently by-products of former feldspar and mica mining activities. The dumps are generally large and easy to locate, and retrieving mineral treasures is usually a straightforward (though time-consuming) operation. Fortunately, the equipment required for dump collecting is nearly identical to that needed for prospecting.

This is how it usually goes:

1. Refrain from pounding on the quarry wall. This has been done for decades by everyone and their uncle. Everything disclosed has either been recovered or destroyed (usually destroyed).

2. Pick a section of the dump that hasn't been dug yet. This will be the most unfriendly place with huge trees and/or poison ivy. Roots, huge rocks, mud, and other obstacles will be present.

3. Dig a hole in the dump's side with your potato rake, shovel, and small camp saw. Make the hole large enough to work in comfortably. Keep an eye out for rocks that may fall on you as you dig.

4. Inspect the rocks you've removed. Look for anything that is brightly colored. Use water to wash the rocks if there are any nearby, such as in the quarry pit.

5. Using your gem screen, screen the dirt. Examine the hue and crystal shape of what's left on your screen. Submerge the entire screen in water and shake to remove the gravel if you have access to water. If there are any jewels, they will be obvious.

6. Continue doing steps 4 and 5 until you are exhausted.

After an hour or so, if you haven't found any good indications of collectible minerals or gems, transfer your operation to another region and start over! Every place is unique, and it usually takes a day or two of ineffective digging to "learn" what to look for. Some portions of the waste will be mineral-poor, while others will be extraordinarily rich.

A wider screen held by ropes from a support beam of some kind, such as a cut-off sapling implanted in the ground at a 45-degree angle, or even a custom-built wooden support frame, is used by more advanced (called "fanatical") collectors.

The multi-bucket wash system can be used if water is within walking distance but not in the local vicinity. Three five-gallon buckets are required. One bucket's bottom should be cut out, leaving a 1" flange. Cut a piece of 14-gauge hardware screen to fit the bottom of the bucket and staple or bolt it to the 1" flange. Fill the remaining buckets with water, one for "coarse" washing and the other for "fine" washing. Pour or shovel the screened gravel from the primary screen into the bottom-screened bucket after screening the dump.

To wash the gravel, immerse it in the coarse wash and shake it, then dunk it in the fine wash and shake it again. Any jewels or fascinating rocks will be seen when you dump the clean, damp gravel onto a level surface. Ten gallons of water will last longer if you use this strategy. A larger shovel will come in handy while executing this type of larger-scale job.

Bring a container for your gems and newspapers to wrap specimens in.

Although we should be delighted that these already operating mines will have new dumps in the future, it is frustrating to know that some of the classic digging places are now off-limits.

After exploring the fascinating history and the motivations behind rockhounding, you're getting closer to diving into the tools and equipment essential for this activity. Rockhounding is not just a rewarding and enjoyable hobby but also a fulfilling way to connect with nature and discover its hidden treasures. However, before you begin, it's important to consider a few crucial aspects and take some necessary precautions to ensure your experience is both safe and enjoyable.

Weather

Know what to expect in terms of weather so you can pack and prepare appropriately. Because snow might obstruct access to a facility, knowing the snow level at a particular place is helpful. Digging for lengthy amounts of time in the heat can be dangerous, so find some shade, have enough water on hand, and schedule breaks in between. People have gone out to the high desert in the summer wearing only shorts, assuming that a desert is naturally hot, only to discover that the desert can get very cold at night. Rain in some regions might make driving to a site unsafe as you could get bogged in mud or slide off the road. It's always better to be safe than sorry.

Animals

There are many different animals in North America and seeing them in the wild is one of the many benefits of rockhounding. Some, however, can be harmful. Even if you don't think you'll ever see a dangerous animal, let alone have an issue with one, being prepared is still a good idea. Please read the following information about various North American animals and what to do if you come across them.

Bears, mountain lions, bobcats, wolves, and coyotes are among the many substantial predatory species in North America. Even though these creatures tend to avoid humans, you can never be too cautious or prepared. Use adequate food storage procedures, especially if you live in an area where bear interactions are common. The smell of food and the availability of food can entice these curious creatures.

Studying the types of wild creatures you might encounter in the area you're visiting is always a good idea. Look up typical predators and what precautions you can take before your journey and in the event of an encounter on the internet.

Always be mindful of your surroundings during your rockhounding expedition. It's easy to become so preoccupied with what's on the ground in front of you that you miss a massive animal lurking nearby. Most animals will avoid you, but if they are injured or sick, or if you get too close to their babies, they may become violent. Always maintain a calm demeanor and work your way out of the situation gradually. You might want to keep a defensive weapon such as bear spray on hand.

Moose, elk, and deer are large nonpredatory creatures that can be harmful. It's best to keep a safe distance from these creatures. They're probably more afraid of you than you are of them, but if they're worried or guarding their young, they can hurt you. So, stay away and simply take pictures from afar.

Rats and squirrels, for example, can spread diseases like hantaviruses, despite being less of a physical hazard. Avoid touching excrement-covered rocks or other items and stay away from any rodents you may come across.

Venomous snakes can be found throughout the continent. Cottonmouths are the most dangerous, but rattlesnakes and the eastern coral snake are also dangerous. Remember that even non-venomous snakes can become aggressive if they feel threatened. While rattlesnakes are easily identified by their distinctive tails, knowing how to distinguish between venomous and nonvenomous snakes and how to treat a snakebite caused by either type is beneficial. In case of a bite, bring any prescribed first aid. Gila monsters and Mexican beaded lizards are both venomous and can be found in the southwestern United States. More information on what to do after a bite can be found online.

Many rockhounds can attest that there are plenty of creepy-crawly insects in every region of North America that can be deadly, unpleasant, or only plain obnoxious. So, keep an eye out and stay safe.

Plants

A rockhound's life might be complicated by a variety of plants found throughout North America. When poison oak, poison ivy, wood nettle, stinging nettle, poison sumac, and ragweed are touched, they cause slight to severe skin irritation. Cacti, brambles, and devil's club can all be unpleasant. If you're probing around near potentially dangerous plants, dress appropriately. Many plants are dangerous, so don't consume anything you find in the woods unless you're sure it's safe.

Tools

Each tool you bring along to aid in your rockhounding adventures has the right way to use it. To avoid any potential malfunctions or injuries, familiarize yourself with your equipment and how it should be handled. Never break rocks with a typical carpenter's hammer, for example. They are not tempered as well as geology hammers and can splinter, scattering metal fragments all over the place. Don't worry; this book has a section that delves more into the specifics of rockhounding tools and equipment.

People

Be aware of your surroundings, whether rockhounding close to home or in a more isolated region. You're not constantly on your own. You may unintentionally come across private property or a dig site that has been claimed by someone else. You must be aware of your surroundings and ensure you are not in danger. Depending on where you plan to dig, you might want to bring along a rock-hounding companion.

Always Be Ready for Anything

Murphy's Law holds true for rockhounding and any other facet of life. Rockhounds tell dozens of anecdotes about bizarre events while in the field. While many of these will make you chuckle, some will leave you with a bad taste in your mouth.

Planning is your best friend when enjoying a fun and safe rockhounding adventure.

TIPS FOR YOUR FIRST ROCKHOUNDING

Remember that your first vacation might not go as planned. It takes time to develop an eye for finding the best specimens. Many rockhounds discover that their second or third visit reveals more of the collection as they get to know the area.

Experience varies depending on location and availability, but finding some rocky areas where specimens naturally settle is always an excellent place to start. For example, it could be under a cliff or on the bank of a river or stream.

Cuttings with clean roadways can also be helpful in looking for specimens. They're conveniently accessible by automobile, making them a fantastic way to climb the cliffs fast without spending the entire day doing it.

Quartz crystals, for example, are invisible to the naked eye. The obvious regions can be found by identifying the roots in the possible areas and then rotating the tool to find them. If you've seen a few, there's a good chance you'll see more in the area.

If you're looking for an excellent gem - or merely a gem, keep the following safety considerations in mind. Time management is also essential, particularly for beginners. Some collectors like to set an alarm clock for themselves to check on the situation now and again. Suppose you haven't seen anything after an hour in one spot; it's better to move on to another. Time might slip by if we are too focused on rockhounding.

If you want to go somewhere else, you can get back in your car and drive away. Going on a few short daily visits is often more convenient than getting them all back at the end. Finally, when going rock hounding, be careful not to overdo it. Stay hydrated and comfortable when necessary. Check that you have the energy to return all of your merchandise and samples.

It's critical to understand the regulations of your new pastime. The rules are divided into two categories: what regions you are allowed to enter and what specimens you are allowed to gather.

A Land Authority can assist you in determining if the area is public or private. Most public locations allow rock hunting, while national monuments and parks do not. Examine the local regulations that apply to the country you want to visit. Trespassing on private property without authorization is illegal and should be avoided at all costs. Everything you steal from private property is the property owner's. If you're unsure, stay away from the area and remain where you are permitted.

The size and type of collections that can be made on public lands are regulated by the Bureau of Land Management. If you have any doubts, contact your local authorities to ensure you are not breaking laws or collecting illegal items. There is a limit to the amount and type of material that can be collected. You should also follow their recommendations for collecting specimens and avoiding potentially dangerous explosives.

You'll notice distinct rules for viewing versus collecting when you look at the criteria for what can and cannot be gathered. Although the two tasks are remarkably similar, the content obtained has a different goal.

Rockhounding is a hobby that involves collecting minerals, stones, and gems for personal use and enjoyment. Demand is the accumulation of something to sell it. Searching for materials on public land is prohibited; however, finding stones is (generally) permitted. Is it illegal to take stones from the natural world? You should be aware. Those fully prepared for their trip and familiar with the territory's rules should not commit this infraction.

Take some time to tidy up and highlight your achievements. If you need to cut or sand the material, investigate which way works best for others. To keep track of how your collection grows, I recommend keeping a journal and writing down everything you find and where you found it. It can also help you figure out what you have without exposing it to the rest of the world.

Most rockhounds enjoy having some standout pieces in their collections. It's more fun to see an impressive specimen than to keep it hidden. I recommend keeping your collection dust-free and giving adequate illumination for optimal results.

From state to state and province to province, the rules and regulations governing where you can travel and what you can do in particular locations vary. It is your obligation to understand what land you are on and the rules that apply to it.

Unless the proprietor gives permission, keep off private property. Fortunately, the majority of accessible rockhounding locations are on public land. These lands are managed by the county, state, or federal governments in the United States. Rockhounding is not permitted on all public grounds in the United States. It is prohibited in some state and county parks, wildlife refuges, national parks, national monuments, tribal lands (unless you are a member of the tribe), and wilderness lands. Check online or phone ahead to see if rockhounding is permitted where you want to go.

Crown land is the name given to public land in Canada, and the provincial government manages it. The laws governing rockhounding on Crown land are similar to those that control much of the federal land in the United States. Minerals, rocks, and some fossils can be collected for personal, recreational, or educational purposes. Only manual collection methods are permitted, and there are strict quantity restrictions. Individual Canadian provinces enforce these restrictions. Before collecting, look up more information about the limits in a specific location on the internet.

The laws governing what you can dig up and how much you can take vary depending on where you are. There are usually limitations on the types of fossils that can be collected. In some areas, there are weight and size restrictions on what rocks, gems, minerals, and fossils you can collect and how much you can collect per day and year. Depending on where you are and what you are collecting, a permit may be required to collect rocks, gems, minerals, or fossils. Always check the laws in the area where you intend to rockhound.

BOOK 1

PRECIOUS GEMS

PRECIOUS GEMS

A cut and polished crystal, mineral, or rock are referred to as a gem. A cut diamond, mineral, crystal, or rock are all examples of gems or gemstones. Amber and pearls are organic materials that are classified as gemstones but are not crystals, minerals, or rocks. Organic matter rather than minerals forms some gemstones. This category includes pearls, amber, and coral. They are used in crystal healing even though they are not technically crystals.

Gemstones are precious or semi-precious stones that are made up of minerals. For a long time, gemstones were believed to be a source of great power and have been part of various legends and myths. Many considered gemstones as sacred treasures that may have incredible healing powers. Some also believed that gemstones improved the user's inner abilities.

These days, gemstones are still considered precious treasures, but mostly because people use them as adornments in accessories and jewelry. They are polished, shaped, and faceted to make jewelry and add value. Some gemstones are extremely hard to procure, and artisans must use specialized tools to polish and refine them even further. This is also why jewelry and accessories adorned with gemstones are expensive and rare. Gemstones have three characteristics that give them distinct qualities: durability, rarity, and overall attractiveness.

The term gemstone often comes to mind when talking about crystals, but there is a difference between a gemstone and a crystal.

Gemstones are precious or semi-precious rock formations used for decoration and jewelry. At the same time, crystals are three-dimensional structures of atoms arranged in a geometric pattern. Most gemstones, such as diamonds, amber, and sapphire, are crystal forms, but not all crystal forms are gemstones. Gemstones, like crystals, can have a mineral base construction or a more organic origin, as seen with amber. It is critical to understand the distinction between the two when purchasing crystals, as you may not be getting what you expect, and gemstones are typically more expensive than crystals.

Some types of gems and gemstones do not follow this strict scientific molecular structure to be considered a type of crystal. These are called polycrystalline. They do not form in the same structures because they are made of more amorphous materials rather than solid repeating patterns. Usually, one of their forms is that of a liquid, such as amber, glass, and opal.

A gemstone is an object that is considered to be valuable because it is hard to come by, has a high level of durability, and is considered to be beautiful. So many of the crystals we will discuss in this book are gemstones; however, some are crystals yet are not gemstones. For instance, as mentioned above, ice and salt are technically crystals; however, they are not gemstones. Because they are not considered beautiful, hard to come by, or durable, they do not count as gemstones. On the flip side, some gemstones are not technically crystals in the molecular sense of their structure. For instance, Coral, turquoise, and Pearl are gemstones, but they are not scientifically crystals. For our book, however, they are crystals.

Gemstones are distinctive sorts of rocks with properties or naturally occurring minerals. The difference between a crystal and a stone is in its atom and molecular construction.

What are Gemstones?

The definition of a gemstone is simple. However, defining exactly what classifies a gemstone is an entirely different animal.

The short definition of a gemstone is a piece of mineral (or another type of rock or organic material) that has been cut, polished, and made into a piece of jewelry. Gemstones are commonly made of minerals, but materials like rocks or amber may also be fashioned into a gemstone. Most gemstones are hard to the touch and high up on Moh's Scale of Hardness, but that is not always the case, as some softer minerals may also be used to create a gemstone. They can be referred to as precious, semi-precious, jewels, fine gems, or simply gems.

Most gemstones are formed in the top layer of the earth - a depth of three to twenty-five miles - which is called the Earth's crust. However, diamond and peridot form two gemstones in the Earth's mantle. The mantle takes up around 80% of the Earth's volume and consists of magma and molten rock. It has a solid upper layer.

The crust is where they are all mined from, despite where they are formed. It comprises three kinds of rock: igneous, metamorphic, and sedimentary rock. They are named after how the rock is formed. Certain gemstones are associated with only one type of rock, while others can be related to more than one.

Igneous rock is solidified magma. The magma oozes out from the mantle and to the crust through volcanoes. Once on the Earth's surface, it is hardened into magma. It can crystallize and create completely different minerals if it does not reach the crust before beginning to cool. High pressure can cause the magma to penetrate surrounding rocks and cause chemical exchanges.

Gemstones formed in igneous rock include:

- Quartz
- Apatite
- Beryl
- Diamond
- Moonstone
- Garnet
- Zircon
- Spinel
- Tanzanite
- Topaz
- Tourmaline

Once this rock reaches the surface, weathering and erosion will create smaller particles of the rock. They will accumulate on the surface and move about by water or wind. Layers of this sediment will build up over time, whether on land or underwater. Pressure will cause compaction in the lower layers of the rock. Once mixed with chemical changes, the **sedimentary rock** will be formed. Evaporation is another that produces sedimentary rock as well.

Gemstones formed with sedimentary rock include:

- Malachite
- Jasper
- Zircon
- Opal

Heat and pressure caused by magma can cause the two aforementioned rock types to change form in their chemistry or crystal structure. This is called Metamorphism. The rocks formed through this care are called, not surprisingly, Metamorphic rocks.

Gemstones formed with metamorphic rock include:

- Jade
- Beryl
- Spinel
- Turquoise
- Sapphire
- Ruby
- Zircon
- Lapis lazuli

As you can see, many gemstones are formed through different rock types, but others are also formed only from one type of rock. All rocks are going through constant states of (often slow-moving) change. This state of change is called the **Rock Cycle**.

Where Do Gems Come From?

Gemstones are extracted from the earth. Mineral deposits from which gemstones can be extracted can be found all over the world, but due to the high cost of mining in first-world countries, they are frequently mined in third-world countries instead. This is not always the case; diamonds are most commonly discovered in South Africa, Australia, and, more recently, Canada.

Colored gemstones are mined in several different places, with the most prominent producers being Brazil, parts of Asia (including Sri Lanka, Thailand, and Burma), and East Africa (including Kenya and Tanzania). Opal deposits are most notably found in Australia and the most critical emerald deposits in Columbia. It depends on a certain environment's pressure, heat, chemical, and rock features. Each gemstone needs a specific combination of these things to be formed, though different ones are spread worldwide.

Gemstones are generally not mined with big industrial machines. Most of the time, they are mined using primitive methods in small teams. The only gemstone that is commonly mined through large industrial methods is diamond.

Because of the various minerals and other elements available, different gems can be found in different regions, but there are certain types of places where crystals tend to grow. Crystals form in rocky areas and underground, especially if these areas are not disturbed. The majority of the earth's bedrock is made up of crystals in one form or another. As previously stated, some crystals form within cooling lava, where most of the world's crystals and gemstones are found. Lava rises from the earth's crust and occasionally pushes through to the surface, forming layers.

The movement of the earth's tectonic plates has a similar effect, which is how new crystals are created. Some crystals are formed immediately as the lava on the surface cools quickly, but others take much longer to form further down, where the lava cools at a much slower pace. The speed at which a crystal is formed usually determines its size.

In some cases, the movement of the lava or earth creates empty spaces where water vapors are condensed into a mineral-rich liquid, eventually growing into different crystals. Many underground caverns are damp and desolate enough that large clusters of crystals can extend for hundreds of years.

In some areas that have mineral-rich earth, it is possible to find small crystals scattered about or in the uppermost layer of the ground. The area will determine the types of crystals that can be found. Individual crystals are more likely than clusters, and many of these crystals will go unnoticed in their raw form. In some sandy areas, it is possible to find crystal formations such as desert roses and sand crystals. There are also a few places where the sand consists of tiny crystals worn down. Quartz is an essential mineral in sand formation, and it is often found that many sand grains are, in fact, quartz crystals.

If you come across a crystal or tumbled stone that you are unable to determine what exact type of crystal it is, there are a few ways that you will be able to narrow down the possibilities of what it is. As discussed in the shapes that crystals take, you should be able to use this information to help determine or cancel out the types of crystals that it is or is not. The patterns present in the crystal, even if tumbled and sanded down to a smooth round shape, are closed.

Beyond that, you can also use the color to help you figure out precisely what crystal you are looking at. Looking at resources such as books, The Internet, and going to a local shop might help you figure out what crystal it is.

When it comes to identifying which crystal will be best for the purposes you seek, that is another process entirely. Numerous crystals can be used for every single goal you can imagine. Therefore, there are many correct answers about which crystal you should use. That being said, we will get into that in detail in another chapter of the book, depending upon what you are looking to use the crystal for.

Crystals and what we predict as gemstones possess a vibration that is free of patterns that are resistant. They are among a number of those constructions from the measurement which have balanced, intentional, strong, and cohesive frequencies.

Their unchanging physical construction reveals that their energetic patterns of strength, balance, and cohesiveness are incorruptible.

In the physical measurement, when you get a crystal, it might look as if you are not doing more than looking or touching another physical thing.

This means that the natural development of vibration would entrain and resonate in the path of health.

Because of this, your electricity will become saturated with the crystal's energy and take on a cohesive pattern when you talk about the distance of a crystal (or bead) with a resistance-free vibration as opposed to the crystal's vibration adopting a non-cohesive pattern. The bodily symptom goes away because you stop performing the disease-causing routine inside your energetic substructures. As a result, the sick energy's physical manifestation disappears. Therefore, we can see how crystals and diamonds are proficient at bringing us back into a state of stability and health.

Anything having an inherent energetic pattern of non-resistance can behave by offering a vibration that we could use to re-tune ourselves into a healthy vibration.
This happens on the energetic levels of the substructure if you listen to a tune that makes you feel good, spend some time near a person that makes you feel good, or take homeopathic treatment.

Every crystal or gemstone resonates energetically with a slightly different pattern and therefore looks different physically in terms of chemical geometry, color, composition, structure, and texture. Because of this, each one lends itself to patterns that live within our systems.

For example, rose quartz would expose the lively patterns active in our physical and metaphoric hearts to align themselves with health and embrace a much more immunity-free pattern. Therefore, when we entrain with increased quarts, unresolved heart problems will dissipate, allowing us to let go of what's distorting our energies.

Gemstones and crystals grow profoundly within the earth's crust over countless years at heat and substantial pressures. This gives them a place among the objects on earth with energy. They're capable of reflecting energy, including casting, emanating, refracting, and receiving.

Crystals have a consistent arrangement of atoms. These atoms, derived from the stone known as "quartz," vibrate at a stable and measurable frequency. As a result, quartz is an excellent receiver and emitter of stored energy. As a result, quartz is used in watches, radios, and a variety of electronic devices.

Just like power, thought is a type of energy that could be given leadership by what we call "intention." Crystals can be programmed without power, utilizing only thoughts as informational energy.

Quartz is also a piezoelectric material, i.e., something which produces an electric charge when stress is applied.

When a conductive substance is placed under mechanical pressure, a shifting of the negative and positive charge centers in the substance takes place, ending in an external electrical field.

The stress can also be caused by twisting or hitting the material without fracturing it to deform its crystal lattice. This effect works the other way with the material when a small electrical current is applied to deforming.

While the argument rages back and forth as to if the piezoelectric effect contributes to the human-crystal recovery relationship or not, the fact that crystals are so responsive to electromagnetic fields has severe consequences.

It has implications since our bodies are composed of and constantly emanating electromagnetic areas, and crystals and gemstones respond to this power that is generating and coursing through our bodies.

Another interesting finding is that quartz is composed of silicon and oxygen (SiO_2), a mixture known to Geologists of minerals as the building block. Our world consists of minerals comprising silicon and oxygen

Silicon is a significant constituent of our bodies. Some scientists have speculated that the transport of energy in the natural crystal into the silicon within our bodies might have something to do with the physical healing effect caused by exposure to crystals.

Crystals and gemstones are just one of the most powerful tools available to people in the physical dimension of presence. It's a tool all people have employed subconsciously at one time or another. Often it's an interaction that occurs by our attention being drawn to a specific stone.

We think it is "pretty," although we do not understand why we like it so much, we feel compelled to pick it up and keep it in our pocket. We've got no idea why we believe this urge, and we don't have any conscious awareness of what is behind our impulse to pick this up.

We have no idea that our lively substructure is calling us towards the particular vibration of that rock to entrain together and proceed towards a more cohesive, healthy pattern than the one we are currently maintaining in ourselves.

Like all tools, the key to utilization is to learn how to implement the tool consciously.

If this type of situation above could cross from a subconscious compulsion to a conscious process of looking for a particular crystal or gemstone based on an understanding of the benefit of using it, the person's receptivity will be that the effects of the entrainment would be a hundredfold.

Suppose we recognized that the quality of gemstone and every crystal impacts our fields. In that case, we'd see that they create an electro-chemical response inside our body and psychology, allowing us to utilize them since they are the tools.

We could use them daily to promote evolution, awareness, growth, and well-being in ourselves and our lives.

COMMON GEMSTONES AND THEIR USES

Amethyst is utilized for problems in the blood and respiration issues. Amethyst crystal clusters are used to maintain the atmosphere and vitality in the home favorable and clean. Amethyst clusters, points, or various tumbled Amethysts laid in a window that absorbs the sun's rays throughout the day are incredibly beneficial in dissipating negativity in the house and in healing. On the other hand, those placed beneath the moonlight will make everybody in the dwelling feel less agitated.

The favorable spiritual feelings will be expanded by utilizing an Amethyst as a meditation center. Amethyst helps defeat anxieties and cravings. It also helps relieve headaches.

When meditating, hold an amethyst stone in every hand. It is a superior stone for better meditations as it allows better visualizations. Set some amethyst stones around where tempers might often be riled, like high-pressure professions and business. It is a rock of serenity and helps bestow love and happiness to all who use it. If you are addicted to anything and are working hard to check the dependency, an amethyst jewel crystal can help. Hold a rock, request the desire to be removed from it, and then draw strength out of the stone. It enables you to do away with a variety of dependence.

An Amethyst stone is a fantastic gift for anybody who works as a psychic as it helps increase all forms of psychic means or those that show psychic powers. Here's an easy crystal healing curative proven to help if you endure migraines. Lie down and close your eyes. Put an amethyst stone on your forehead, attempt to unwind, and allow the gemstone to do its work. By setting an amethyst stone inside an elastic bandage that has been wound around the wounded area, historically, muscle and joint traumas like sprains have been helped to cure quicker. To fix breathing issues faster and get medications from a doctor, put an amethyst between the lungs and the torso.

Depending on the severity of the illness, you may be able to tape a jewel with a band-aid and sleep with it in place. To make an amethyst stone elixir, set one or more amethysts into a transparent glass jar full of water. Let the water sit outside in the moonlight for the whole night. The closer to the full moon, the better. You might be able to use amethyst water to help clear up blemishes and soften the skin. You use it as an ingredient in virtually any clays or masks you might implement or may wash with it.

Wear an amethyst stone around your neck, taking one in your pocket to help fortify your bones. If you spend more hours tossing and turning than extremely sleeping and find yourself having issues sleeping at night, place an amethyst jewel beneath your pillow to assist with sleeplessness. When you awaken, utilize an elastic hair band as a headband around your forehead to enlarge some dreams which you have and to help you recall your dreams. Slip an amethyst stone beneath the band as it is proven to help you dream. Bury a little amethyst stone at each entry to your house to guard against robbers.

An inexpensive strand of amethyst chips is ideal for this. If you have a window above a cement patio or veranda, bury a small amethyst jewel crystal formation, and remember to bury a little beneath the doorways and every window. To keep evil out of your home and all who wish you harm at bay, use the same technique described above for your loved ones or as a shield against thieves. If a person is looking for a suitable partner, someone who will work with him on a journey to create a life together, advise them to keep an amethyst rock in their pocket.

If you're a woman and believe your man could be losing interest in you, give him an amethyst stone as a present to enlarge his attraction to you. This might take the form of a pendant or a ring. Even a worry rock in his pocket will do fine.

To commune using your Higher Self or your Spirit Guide, choose a peaceful time and place where you will not be interrupted. Carry an amethyst rock in each hand. Take a couple of deep breaths, shut your eyes feel the powers come from the amethyst. Enable them to come up your arm and into your head, where you see them begin to burn from inside your mind's eye. Invite your guide to come forward and speak with you. This can help you attune with your higher self.

Make an Amethyst Jewel elixir and utilize it to bathe the portions of the body which are getting circulatory problems. It encourages circulation in both the etheric along with the physical body. With the angry frenzied world moving so quickly around us, we often find ourselves extended beyond the ability the human body can take. Spend a couple of minutes consuming the energy of amethyst crystals to mend the nervous system.

You have probably heard it being said before: "You are your worst enemy." Self-deceit, especially concerning heart matters, induce chest pains and more heartaches than humankind has ever mustered up. Amethyst stone crystals protect against self-deceit and let you view things as they are.

Amethyst is well known for working with the third eye and is strongly associated with and supported for work with the third eye for those interested in chakra work. It is frequently used in chakra work when exploring higher, spiritual realms because it is a defensive stone against spiritual attacks and negative psychic influences that could harm your career.

This translates well to meditation if you're not necessarily doing chakra work to open yourself up. This stone is a must-have if you are more interested in spiritual exploration. This stone is renowned for its protection.

Those seeking the higher properties of Amethyst in their daily lives should expect tranquility, peace of mind and heart, and, most importantly, sobriety when carrying it. The legend of Amethyst's formation was linked to the lament of the god Bacchus pouring out his wine of the crystallized maiden of unparalleled beauty, Amethyst, by the Romans. This stone will help you recover from a hangover and will protect your mind from psychic invasions that are all around us.

Apophyllite

Apophyllite is an excellent rock for calming an overactive mind due to its frosty white and cool blue tints. It can also aid in preventing overly energetic destructive behaviors such as nervous twitches and tics.

On the other hand, Apophyllite is a rock that is as much about avoiding problems as it is about curing them. It works by bringing to light the underlying causes of specific behaviors and harmful habits. It is not always a pleasant procedure, but it is necessary for spiritual development and new stages of soul development. This stone, too, aids in easing the tension of these transitions.

Apophyllite is abundant in Australia, Brazil, the United Kingdom, the Czech Republic, India, and Italy. It can be transparent or opaque; however, translucent apophyllite is more uncommon. Apophyllite is commonly colorless in fine crystals but can also be green, peach, white, or yellowish tinge. It usually takes the shape of a cube or pyramid shape and produces microscopic crystals. However, it grows in abundance, and enormous apophyllite clusters are typical.

This crystal is a good conductor and strongly connects to the element water. It can magnify the energy within a place and transfers vibrations well, making it ideal for amplifying the skills of other crystals. It increases clear vision and intuition while grounding you to your physical body during deep meditation or out-of-body experiences.

Apophyllite enables you to examine your inner thoughts and emotions and assists you in correcting any errors or imbalances you discover. It motivates you to tell the truth to yourself and others. It is a soothing stone that aids in alleviating stress and removing mental barriers. It aligns the mind with the spirit, increases compassion and love in decision-making, and decreases craving.

This crystal aids with the reduction of anxiety, the alleviation of fear, and the tolerance of ambiguity. Also, it is beneficial for various methods of energy-based therapy, such as reiki, because it induces profound relaxation and improves energy transmission. It is also advantageous to the skin and sinuses, aids in allergy relief, and can be held against the chest to prevent asthma attacks. It can help cure the spirit and revitalize the eyes by placing a little apophyllite stone on each eyelid for a few minutes.

Obsidian

When lava cools too quickly to crystallize, obsidian is formed. It's found all across the world, but mainly in Mexico, and while some hues are familiar, others aren't. Some colors, such as blue-green, are nothing more than shaped and processed glass.

It's a translucent stone with a highly reflective, glass-like appearance. It comes in various sizes in its natural state, but it's also available as a tumbled stone. Its colors are black, blue, brown, green, red-black, or silver. It can also have a golden shine or multiple hues on one stone, known as a rainbow.

Obsidian is a highly effective stone that acts quickly. It promotes honesty and effectively brings flaws and obstructions to light. This stone encourages you to develop as a person while also supporting you. Because it has such a tremendous ability to reveal terrible realities and negative feelings, it should be handled with caution and under the guidance of a specialist.

Obsidian can also bring about deep soul healing. It's a protective stone that may ward against negativity and is beneficial for anchoring and developing the root chakra. It can also shield you from psychological and spiritual negativity. Obsidian is a stone many people find overwhelming, but it is a vital tool for therapists.

Obsidian is a fantastic stress reliever since it has a relaxing impact and can help you relax. It also assists in confronting your issues and finding a long-term solution to them—obsidian aids in removing ambiguity and establishing clarity. Because of the bad energy it absorbs, this gem needs to be cleansed frequently. It helps remove blockages, tension, and toxins from the body, improve digestion and blood circulation, and relieve pain from cramps, troublesome joints, injuries, and arthritis. It has the potential to help people with clogged arteries and an enlarged prostate.

Onyx

This stone is widely available in Brazil, Italy, Mexico, Russia, South Africa, and the United States.

It's a ringed stone that has a marble-like appearance. It's often marketed as a polished stone in a variety of sizes. Its primary colors are black, blue, brown, gray, red, white, and yellow.

Onyx is a reassuring stone that helps strengthen you in times of stress, confusion, and difficulty. It can help you focus your energies, increase your vigor and stamina, and strengthen your resolve. It makes you feel more at ease in your surroundings and boosts your self-esteem. It works wonders for old injuries, trauma, and grief. It can assist you in overcoming fear and concern, making better judgments, and improving your self-control. Obsidian treats problems with the blood, bones, teeth, bone marrow, and foot by strengthening your body's inherent ability to heal.

Aquamarine

Aquamarine, as the name suggests, is a rock with water, beauty, and nutritional properties. You may believe that the core of existence awaits you in this rock, which is so powerful that its cleansing powers are frequently invoked.

But this is also a rock about the tides of transition in life. When you have difficulty getting away from challenging circumstances and interactions, Aquamarine connects you to your internal power. It can also assist you in positioning yourself with other individuals on the same trip.

Aquamarine is a well-known gemstone found in Afghanistan, Brazil, India, Ireland, Mexico, Pakistan, Russia, the United States, and Zimbabwe, among other places.

Another crystal with a range of transparency to opaqueness. It is commonly marketed as a tumbled or faceted raw stone with a blue-green tint.

Aquamarine is a stone that helps calm the mind, provides calm bravery, and relieves stress. It also aids in pollution protection. This stone has a long history of attracting the attention of light spirits and being used by sailors as a charm to protect them from drowning. Aquamarine is an excellent stone for those who are sensitive and for those who are feeling overwhelmed. It can assist you in becoming more tolerant and less judgmental. It also inspires people to take charge of their own lives.

Aquamarine is excellent for clearing your mind of unwanted rogue thoughts, sharpening your mind, and resolving any uncertainty. It promotes openness, self-expression, and closure. This stone aids in the healing of the throat chakra and the development of clairvoyance. It's also a good stone to use while you're trying to relax. Aquamarine is an excellent immune booster, but it is especially beneficial for throat issues. It aids in strengthening the eyes, stomach, and teeth, as well as treating fundamental immunological problems such as hay fever.

Aventurine

Aventurine is a stone tightly linked to the Heart Chakra, where some ancient sayings about pursuing one's soul are born. Harmonizing with the adventurous vibrating nature of this crystal, these ideas encourage you to follow your heart with boldness!

There is a sense of playful risk involved with Aventurine. Many have named it the Gambler's Stone, thinking that it links you with excellent luck and fortunate spots. While every individual perceives fate differently, letting a little opportunity come your way hardly hurts!

Rutilated Quartz

Rutilated Quartz is appreciated by many faith healers and spiritually minded crystal enthusiasts because of how adversities can be so efficiently filtered out of your life. If you find yourself defeated by pessimism in everything around you, this crystal can restore that view to a more balanced one.

It can also assist you in cleaning up your negativity, too. It can cleanse your soul and mind from any anger you have, even if it occurs in your subconscious. Many of these adverse feelings can slow down your religious development, making this stone a precious ally.

Selenite

The nature of life and warmth makes Selenite a vital crystal for anyone who wants to see the spiritual truth. It is known for helping to open your third eye Chakra; because of that, it can direct you to intriguing spiritual ideas.

It's a purification crystal for yourself and your surroundings—many place Selenite in their home or office to keep negative emotions and small vibrational energies at bay. You are frequently ready to be your most fabulous self by keeping yourself light and airy.

Serpentine

Serpentine is a rock that operates with the lesser Chakras and, as such, can assist in easing any concern you have with the physical globe—or with physical pleasure.

If you are so anchored in the physical world that your greater and spiritual self is entirely out of equilibrium, this crystal may even be a bit of a scale. And, of course, its color helps awaken you to the curing power of nature and how you can take greater care of yourself.

Serpentine is a crystal that helps to bring wealth in all its types, making it a good luck charm for many individuals.

Jade

Jade is a mystical and powerful stone that will be familiar to anyone looking to work on their heart chakra. It is known for drawing and brightening all heart-related issues that come to a person, whether it is love or loss. Those looking to do extensive heart chakra work might be interested in trying jade, though it is relatively uncommon among those working with the heart chakra. However, its popularity is on the rise.

For those looking to work with love or solve issues involving dreams, you may want to consider meditating and working with jade. Use jade if you're looking for a strong stone to help you solve the riddles of what you might have been troubled with in your sleep. Its powers link it to the spiritual realm and will help you discern the information that the Divine or your higher self is looking to impart subtly to you. It's a way to predict coming storms. As for those seeking questions of the heart, jade will help you find your answers.

For those looking to carry jade, it is a stone that will help you with your heart and the matters of love in your life, romantic or respectful. For those looking to draw trust and loyalty towards them, jade will help you. If you're looking to spend time with those you love and honestly care about, you may want to consider putting a jade in your pockets or wearing a piece of jade jewelry.

Jasper

Jasper is often called the nurturing stone, and this quality is what keeps it in many meditative households. It is a stone that keeps its user calm, relaxed, and safe. For those looking to maximize the benefits of this crystal with their chakra work, consider utilizing it around the navel and heart chakra. It's a powerful grounding stone and particularly good at cleansing auras.

Because it is so good at cleansing auras, jasper is often used at the beginning of meditational times and is excellent for grounding you after meditation. For the actual meditation experience, Jasper has a slow, steady vibration that will help you find the stability and comfort you need when facing large problems. Jasper is there to protect you and guide you safely through the scarier moments in your life, so embrace its use.

Also, Jasper is known to have a solid tie to beauty, which means beauty is drawn to jasper. You may notice that using it in your meditation makes you experience the beautiful moments in life instead of the scary ones. That's okay; it's all part of the experience to find out what you need to learn.

Carrying jasper is an intelligent move, and it is excellent at helping you deal with troublesome situations that may arise. But more importantly, use jasper intentionally. Since it is grounding, you'll have a calming sensation best targeted at those terrifying days. Again, remember the element of beauty that is drawn to jasper. You might find yourself surrounded by awe-inspiring sights and beautiful people.

Moonstone

Moonstone is a powerful stone in many colors that can be selected to align your chakras, but it is most notably used with the solar plexus. This is a powerful stone, but you can find many variations that will let you use it more effectively based on the chakra work you are trying to perform.

As far as meditation is concerned, moonstone is most effective when following the moon's path. At a full moon, you will find yourself traversing into deeper levels of insight and understanding. When the moon is full and new, you'll want to utilize moonstone so that you can find true insight regarding the problems you are facing. It will help guide you there, especially if you decide to travel.

Moonstone is known as the Traveler's Stone, and it is perfect for carrying with you as you go about your business. It will take you to new places, encouraging you to wander and explore. Remember that as you go, your insight and awareness will be heightened by the presence of the stone. This makes it perfect for exploring new places on trips or vacations. Let the stone work through you, and you will find new things to broaden your horizons.

Peridot is one of the more exciting stones and a vital companion for healing and discovering new possibilities through rebirth. Regarding chakra work, peridot is most effective in the solar plexus and will help you most efficiently there. Use it wisely and find out the benefits of having peridot in your life.

As far as meditation goes, peridot is a stone of rebirth. It is commonly discussed how it is most effective with those beginning new journeys and traveling, spiritually or physically. For those reborn, the peridot will guide you and help along the way. This is powerful in helping you bolster your awareness and psychic prowess at the beginning of your enlightenment.

For those looking to carry Peridot on their travels, you have chosen a perfect stone for you as a companion. Peridot will help you with your awareness and insight into new situations. Remember that the energy of the Peridot will cause your aura to desire new experiences, to help define yourself in a new light, and most of all, it is looking to help you see the world differently. Keep it with you as you go on your journeys, and you will discover newer and more complex truths about yourself.

Chrysocolla

Chrysocolla is a beautiful piece for chakra work because it cleanses an area and utilizes and activates all of the chakras. Those looking to contact the divine and genuinely communicate with the universe are enormous fans of Chrysocolla because it is a straight-to-business stone.

This is great for meditational purposes because whether you do chakra work or not, your chakra directly influences your meditation. When your chakras open and there is a sense of equilibrium through the body, you will find yourself in harmony, banishing the negative energy from your soul, and meditational qualities will enhance and speed up. You will find your mind open, and distractions will be all but gone, paving the way for you to seek the answers to the questions that are on your mind.

As far as carrying this stone goes, Chrysocolla is all about the harmony of the soul and activating the chakras. This means you will experience all of the positive effects of harmony. This means you will have an overall feeling and presence of tranquility in your life. You'll not be bogged down by negativity and feel a banishment of fear, anxiety, regret, and guilt. You'll find yourself filled with positive emotions that won't be constrained by the harmful smog hanging around your soul.

Flourite

Fluorite is a famous stone for meditation and chakra work to cleanse auras and dispel negative energy. Fluorite is often said to be one of the stronger crystals in helping organize your thoughts when meditating and trying to focus. This is a stone that allows for regulating spiritual order. That order will help show up in your meditations by offering you a sense of clarity and calm to work.

Carrying Fluorite with you will help breed the benefits of having your spiritual essence in order and your chakras aligned. That means you will have boosts in focus, mental intuition,

and understanding of subjects. It will give you the boost you need to promote stability and have a clear head to think properly.

Garnet

The garnets have a special, enormous ability to protect you from evil and are all about the base, sacral, and heart chakras. Everything revolves around psychological defense, especially while moving physically.

This is a stone to take with you while meditating and on your day-to-day adventures. By keeping garnets with you, you can keep yourself protected from hostile, external powers.

As a darker stone, Garnet is intense with keeping you grounded and from wandering. This stone is great to take about if you're going through a crisis or depression because it's a powerful stone that expels negativity and shields the mind from harmful energies. It will help you find your way through the darkness, grounding your thoughts in real solutions.

Hematite

For those looking for a stone that will help them ground themselves after they've spent a lot of time meditating, then Hematite will be what you want to acquire. This is undisputedly the most common grounding stone that stabilizes and protects those working with it. It's a strong stone tied to the Earth plane. It's also known as the stone that helps remember spiritual journeys the most. Because of this grounding nature of Hematite, most people associate it with root chakra work.

For those who wish to carry Hematite around daily, you'll want to be aware of the grounding effects of this stone. You will be calmer and more in tune with the world overall. It will help you with the logical side of your mind, like retaining numbers and thinking of solutions. This harmony will most likely help you boost your confidence and intellect.

This powerful stone will help keep bad energy from affecting your daily habits. If you're going to be influenced by power and want to keep an open mind that isn't overly distracted or pulled by other crystals or energy, consider using Hematite.

Howlite

When it comes to howlite, it is mainly associated with the third eye chakra and is pretty much held there with much respect. Of course, you're welcome to experiment with other chakra work, but most results are found with the third eye. The reason may be different from person to person, but the consensus is that howlite is a calming stone, regardless of the color that it is found, which is rare. Most stones vary in effectiveness based on their color, but not howlite.

For those looking to work howlite into their meditation, it's an amazingly simple stone to incorporate, but it is dominant in exploring the spiritual realm. This is due to the stone's incredible calming effects that decrease fears, worries, anxieties, and selfishness. This, of course, helps those working with this stone to find concerns and truths that might be hidden very deep inside of them. This is where we start talking about past life work. Howlite is an extremely popular stone for those looking to explore and find answers to their past lives.

Those carrying howlite with them in their travels can often express a calming sensation that focuses on their worries and anxieties, which transfers your energies into something more practical, such as communication. So, expect heightened communication ability and calmness from howlite; you might even start to experience a rush of creativity flowing from past lives.

Lapis Lazuli

Lapis Lazuli is a powerful stone close to those who like to work with their chakras to develop spiritually. As with all blue stones, lapis lazuli has a strong bond to the throat chakra, but it's not limited to that. It also has strong ties to the third eye and the fifth chakras. This versatile stone will help you utilize it in multiple ways to get a lot of chakra work done with a single stone.

For those that like to use lapis lazuli in their meditation, you will find that lapis lazuli has a rich and influential history in the world, and that's for a reason. Kings,

Pharaohs, and Emperors have all been drawn to lapis lazuli because of its allure towards leadership, alignment, truth, wisdom, and growth. In your meditation, notice that you will find yourself developing with lapis lazuli. It is an active stone seeking to make changes and develop your leadership, courage, and harmony. You want to utilize this stone if you need work and progress in your personal development.

Carrying lapis lazuli with you in your day-to-day interactions will bring you to the alignment of truth and wisdom, giving you an extra push in all things surrounding leadership. If you are looking to make intellectual pursuits, then you will want the guidance and wisdom that is offered with lapis lazuli to help you on your path. This is a powerful and versatile stone that I would suggest everyone getting into crystals acquire. There's just so much that you can do with it.

Malachite

You can do many things with Malachite regarding chakra work, but experts mostly use it for the heart and the solar plexus work. The reason is that solid protective powers come with malachite that helps unveil things and pull away from the negative energy in those areas. This stone helps absorb that energy, so it must regularly be cleaned.

As far as meditation goes, malachite has an extensive history of combating negative energy. For those who feel that they are being bombarded with negative thoughts or draining tasks, this stone will help you keep your thoughts clear and focused on what you need to think about. This is a powerful stone that will help you. But as was previously mentioned, Malachite absorbs negative energy and doesn't dispel it. So, you will need to clean it when you decide to use it regularly.

As far as carrying malachite with you, it will help protect you from harmful waves, dangerous energy, and any kind of psychic attacks you might encounter along the way. Malachite is very protective, but if you notice that you're starting to feel less and less of the effects, you need to cleanse it. So, take the time to monitor your malachite carefully.

Apatite

Before you might be in a position to change something, it has frequently been said that you must accept things as they are. Apatite is a multifaceted gemstone that helps us connect to our inner selves and take on the balancing, communication, healing, and teaching it needs to provide. It is suitable for people whose issue is excessive weight.

Apatite gemstone is perfect for appetite suppression. It does not exclusively lessen your appetite. Instead, it lets you see the truth inside yourself so that you can get to the root of overindulging, which is an issue that must be solved in pursuit of weight loss.

It is the perfect gemstone for utilization on any chakras as it could both perk up under action and calm down over action and clear congestions in almost any of the chakras. Apatite helps you attune your brain, heart, and soul to the religious forces that run through the entire universe and guide the evolution of psychic powers. It also helps bones mend quicker and stronger. It assists your body in absorbing calcium from your foods, which helps keep teeth and bones firm.

To help alleviate the pain caused by arthritis, wrap the involved joint within an elastic bandage, letting it hold one or more jewels from the impacted joint. The apatite gemstone can help fix the joint quicker and cure the painful sensation. One or more apatite gemstones can be used to create an elixir. Leave the mixture outside overnight, preferably under a full moon. Drinking this elixir can help strengthen bones and treat and prevent joint discomfort.

To help lower hypertension, wear an apatite gemstone in a way that it hangs just above your heart. You can pin the stone to the inside of your top.

If you are aware of your propensity to let instinct override reason, especially in emergency situations, apatite gemstone crystals may be your answer. This gemstone will promote serenity, giving you the power and time to let reason take control of the situation and produce some creative work. Thanks to it, you can connect with your true center and produce amazing work.

Do shyness or self-doubt prevent you from enjoying yourself at events or in recently introduced social situations? An apatite gemstone may give you the self-assurance and security you need to shine at your best and socialize with people.

Meditate with an apatite gemstone resting against your third eye chakras (roughly above and between the eyebrows) to increase your ability for future visions. The best apatite stone for this is blue or purple in hue.

Do you need a bit more inspiration to complete the job? During meditation, holding a reddish apatite gemstone or a bar of gold may help you keep your thoughts on the task at hand and inspire you to work until it is finished.

Green Serpentine

Serpentine is an earth jewel that helps meditation and spiritual exploration. It opens up psychic abilities and energizes the crown chakra, clears up the chakras, and helps us comprehend the religious basis of life.

This stone's assistance in retrieving wisdom opens new pathways for Kundalini energy to grow up and recover memories of previous lives. Serpentine assists you to be more in command of your life, corrects mental and psychological instabilities, and helps the conscious direction of healing power towards troubled places.

Physically, a Serpentine mineral is detoxifying and exceptionally cleansing for the body and blood to ensure longevity. It does away with magnesium absorption and parasites, aids calcium, and treats diabetes and hypoglycemia.

Light-Green Serpentine is a gentle, tender-natured stone that gets you into contact with angelic guidance. It gets at and integrates the previous, present, and future and is an excellent stone for the past-life quest. This stone supports compassion and forgiveness for yourself. Holding this stone directs you into the healing regions that exist in the between-lives state to ensure healing that wasn't undertaken after a former life stopped may be realized. It removes emotional baggage from old relationships and instabilities and heals from past lives. It can help speak of the past and concludes issues carried forward into the here and now if put on the Throat. The stone is lovely to use when you need to confront anybody out of your past, as it brings a gentle touch to the assembly. Light Green Serpentine is fantastic for irritations, particularly menstrual and muscular aches and pain relief.

Turquoise

Turquoise gemstone is the healing stone that attunes our physical selves to the biggest lands. It helps us better understand ourselves and bring emotions and our thoughts under command so we might see them get fruitful in our reality. You must stop and listen, be quiet and prepare yourself to hear the truth about whom and what you are. Simply then, are you going to find your full power?

Valued by the Native Americans as holy, the turquoise gemstone soaks negativity, transmuting it

into valuable energy. It similarly allows you to become one together with the cosmos. The real Turquoise significance comes from the soul of the individual using it and also the heart. The listing of turquoise gem fixing aspects is long and wide-ranging along with the ranges of sizes turquoise crystal shapes, as well as colors which could be used, are as wide-ranging as the people that use them. Worn anyplace about the body, a turquoise gem healing jewel will protect and bless the wearer.

It's considered a sacred stone in some cultures, personifying a gift. You are able enough to align your chakras by setting a turquoise jewel on all of the chakra points for 3 to 5 minutes while the gem executes its work. It might take a trifle longer; nonetheless, setting a single rock will still align your chakras for the level best electricity, should you don't have seven turquoise stones. A strand of turquoise beads used as an anklet, necklace, or bracelet can help the body from radiation, pollution, toxins, and alcoholic beverages. The idea would be to put on a circle of beads around one region of the body so as the blood flows backward and forward through the region, it may purify. So that it hangs instantaneously within the area, inducing the issue, anybody with problems using their lungs, throat, or asthma may hang a turquoise gemstone from a cord or string. It will help the jewel energies start the healing work even quicker and get as close as possible to the trouble area.

Those with depression may sleep with a turquoise gemstone to lift the disorder faster. Add a couple of turquoise crystals into a container of water and let it sit outside overnight where the moon may shine on it and leave it so the sun may shine on it during the day. The following evening, pour the turquoise water right into a tub of bath water, jump in, sit back, and allow the healing energies to work on your body. This same curing elixir might be utilized to soak a sprained or pulled muscle. It fortifies the body so that you can fend against viruses and infections and aid in the healing of cut or damaged tissues. When you have a headache, place a towel that has been soaked in the elixir on your brow until the pain goes away.

Have a date the following day? Sleep with a turquoise crystal close to your neck at night, and wear one on a chain around your neck throughout the day to ensure your power to communicate with others properly.

Quartz

In the pecking order of crystals used in spiritual healing, nothing stands higher up than Quartz. The healing energies of quartz have consistently been recognized. Quartz hasn't been valued more than any other stone since the fabled Atlantis due to its crystal healing properties.

Reiki Masters, Shamans, acupuncturists who use needles coated with quartz, and even science realizes the amazing and unparalleled abilities of Quartz Crystals. The crystalline structure of quartz carries electricity and radio frequencies. It is why Quartz is utilized in added electronic devices and radios and why men of science are experimenting with extra crystals and Quartz as sources of alternative energy that is perhaps infinite.

The "Dilithium" Crystals that powered the Starship Enterprise and the nearly magical crystals that served as the foundation of Kryptonian science in the Superman films were most likely inspired by quartz's true energy transmutation powers crystals. There are many different sorts of Quartz Crystals, and each has its healing powers that are unequaled and impact other parts of the body and help with various ailments. For instance, crystal healers for managing heart issues, headaches, and kidney disease utilize Rose Quartz. Although Clear Quartz is used to draw pain out, bring clarity of consciousness back, and amplify all curative energies.

Nonetheless, all Quartz Crystals can harmonize with and realign the body's vibrations, bringing back equilibrium. That's what makes quartz crystals so efficient in healing. Most illnesses, mainly mental disorders and neurological issues, may be linked to some "chemical" or "Neurotransmitter" instability. The sway of quartz crystals may mend these imbalances. The electro-magnetic aspects of quartz are primarily due to the base of its crystalline anatomical structure being made up of Silica. Silica is a natural glass that is occurring. Silica is detected in some amount in ritualistic gems, Chakra stones, or nearly every healing crystal.

Silica shares its chemical and molecular building with Silicon, similarly famous for its electro-magnetic aspects (does the name Silicon Valley mean anything?) The same basic component people rely on to convey all of this information worldwide and enable you to examine this page might also help us tap into the fabric of the cardinal energies of the universe and communicate. Spirituality and science aren't so aloof after all. Einstein said, "The more I come to comprehend the universe, the more I am convinced of the existence of superior reasoning energy. There are two ways to live: you can live as if nothing is a miracle, or you're able to live as if everything is a wonder."

Blood Stone

Previously, the bloodstone crystal was called Heliotrope. The word heliotrope is compiled of the Greek word for "the sun," Helios, and the Greek word for "to turn," trypsin. The jewel was first used historically to alter weather conditions. When the rock is left in the water, it is rumored to absorb the sun's rays and trigger a thunderstorm. During the Middle Ages, the red spots on the stones were thought to be the blood of Jesus, and also, the early Christians believed the stone held all the powers of Jesus, for example, a capacity to make the wearer unseeable.

Few believe that while Jesus was on the cross, his blood fell at the foot of the cross on some Jasper on the ground. This is how the bloodstone was formed. It was said that the stones from that area had a tremendous ability to heal almost anything, and the gemstone was given the nickname "The Martyr's Stone." A hero's gemstone is commonly seen in armor, breastplates, and swords for its ability to inspire bravery in dangerous situations. Likewise considered to be a powerful fixing stone enabling the wearer to stay impregnable in battle after those around you have fallen.

It improves creative thinking, self-expression, and artistry. In the Middle Ages, bloodstone was ground into a powder, combined with honey and eggs, and given to patients to heal tumors. Spread made of mashed bloodstone and honey was rubbed on wounds to stop the surplus bleeding. To help remedy snakebite, affix a bloodstone to draw the poison out of the bite. Notice: This was an ancient utilization of the stone. I may do this while on my way to get medical help, but to do this rather than getting medical help will likely be irrational.

The ancient Babylonians utilized bloodstone that was engraved in divinations. They used the manner the various spots of crimson appeared to tap into their psychic abilities, creating an impact like a vision by following the collection of the spots. To purge the mind, soul, and body in the nighttime of the full moon phase, find an area outside where you might lay under the moon's light. Place a rock on your brow so that as you lay there, visualize the moon's power entering your body, filling it with perfect white light. As the body fills, see all of the negativism, sickness, and stress leaving from the body, leaving from the rear of the body and sinking to the floor under you.

Ancient Egyptians utilized magic that was bloodstone to assist them in battles. They used charming, empowered rocks as amulets for the warriors to expand their strength. To turn "invisible" for your foes, wear or take a bloodstone and visualize a cloak of power emanating from the rock and enfolding around you, making you unseeable to those you do not need to see. Athletes may utilize a bloodstone amulet to help expand their strength and speed. Wear or carry a rock and visualize its power entering the body and inducing your muscles to become firmer (or quicker).

Anybody needing bravery might use this same magic to get through a situation. Just envision the powers and aspects you require going into your body. Should you know somebody that tends to be a bit too "me" oriented, give them a heliotrope as a present. It enables them to understand how problems affect not just themselves but also those around them and the entire world.

Hold a Bloodstone in your hands with meditations designed to help you associate with your preceding lives. Turn your ideas backward to a time before your birth once you've entered the meditative state, and let the pictures lead you to sights of your prior lives. Keep more or one bloodstone on the work table or desk to help significantly expand your business and riches.

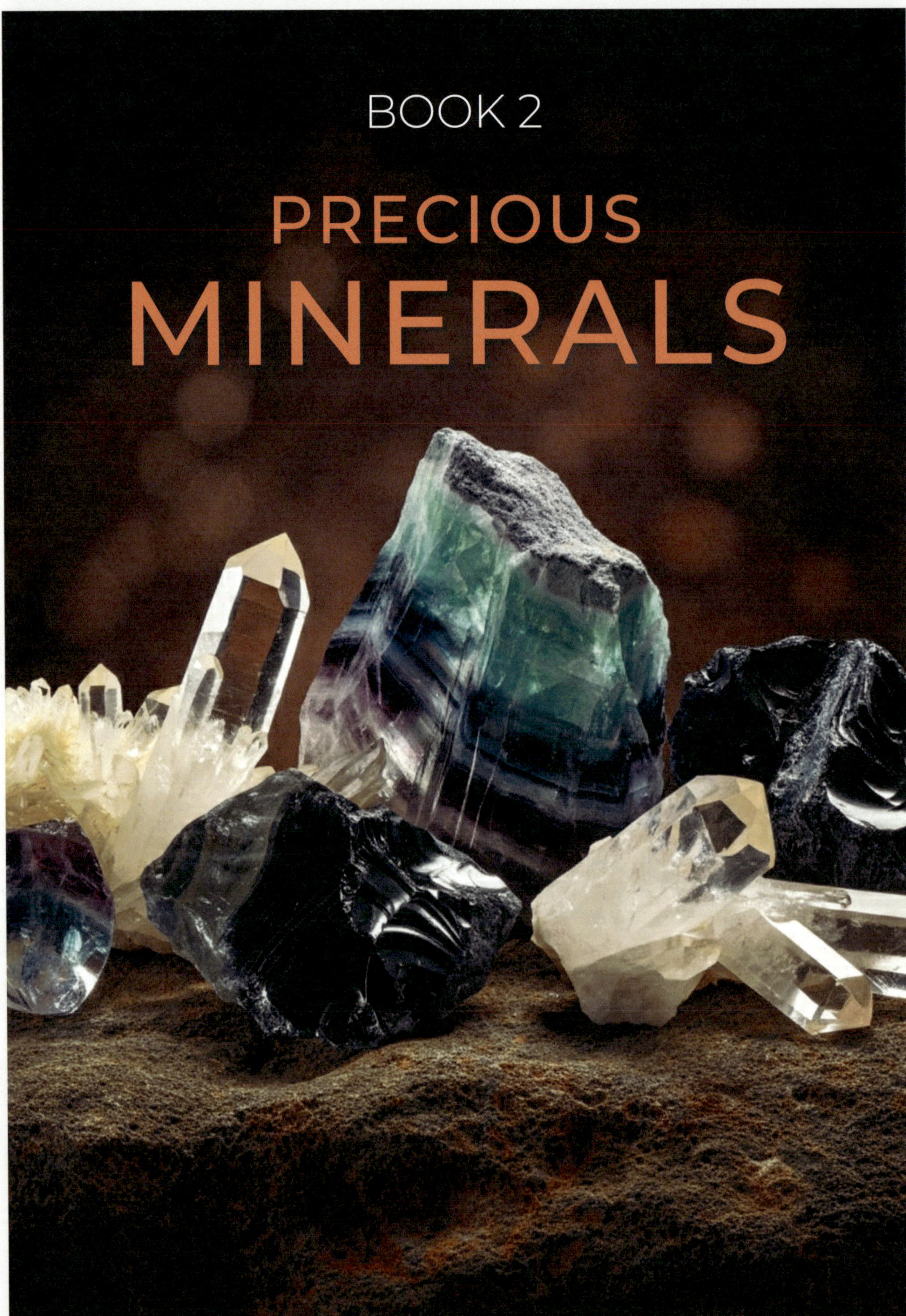

BOOK 2
PRECIOUS
MINERALS

The popular belief about five centuries before Christ's birth was that all matter was comprised of four main elements: earth, air, fire, and water. However, a few great minds of the time, such as Democritus and Aristotle, protested against this popular but unproven idea. They investigated by monitoring a variety of chemicals in various containers at various temperatures. They determined that matter comprises of particles too small to see that shift and move under changing conditions. This was thought of after watching a substance's shape and appearance vary with variations in temperature and pressure—becoming solid, then liquid, then gas.

Few people believed that matter was made up of practically invisible particles for hundreds of years after the Greek tests. However, by the 1600s, most scientists had agreed that all substances are made of a basic particle. This atom has a nucleus surrounded by subatomic particles, or electrons, as a result of experiments testing the theories of physicists such as Sir Isaac Newton, Christian Huygens, John Dalton, and others.

People believed that all matter was made up of different earth, air, fire, and water mixes.

Each atom has a core nucleus surrounded by one or more electrons and comprises protons and neutrons. Hydrogen is the simplest atom, with a single proton nucleus orbited by one electron. More orbiting electrons are seen in substances with more complicated structures.

Copper has an isometric (cubic) structure, which means that only copper atoms are grouped in an orderly cube shape. Each cube contains a copper atom at each corner and in the center of each side, or face, which, when combined, forms a mineral. Extra facets at each corner of the cube might give it fourteen sides instead of six.

Atoms fill an organized structure termed a crystal in a mineral, the solid phase of a nonliving, naturally occurring substance. The crystal may be built up of molecules made up of multiple types of atoms, or it may be made up of one kind of atom. Single-atom minerals include gold, copper, and silver, examples of native elements. The native element gold is made up entirely of gold atoms, while copper is made up entirely of copper atoms. Pyrite is a mineral made up of molecules that looks like gold. Iron and sulfur atoms coexist in their cubic crystal. Pyrite is the mineral name, and iron disulfide is the chemical compound (molecular) name. Table salt comprises the atoms sodium and chlorine, which are kept together by mutual attraction. The chemical name for salt is sodium chloride, and the mineral name for salt is halite.

People fashioned tools from stones or bones before the discovery of metallic native elements for tool manufacturing. Stone Age toolmakers crushed down bone tips to shape spear points and knocked chips from stones to sharpen edges for scraping.

Between 3000 and 1185 B.C., bronze, a mixture of the metals copper and tin, became widespread (the Bronze Age). Around 1185 B.C., when early people discovered how to make tools and weapons out of the element iron, bronze fell out of favor, and the Iron Age started.

The iron and sulfur atoms in the mineral pyrite are organized in a cubic pattern.

On its flat surfaces, pyrite is frequently found in cubic shape, with striations or parallel lines of grooves. Pyrite is sometimes mistaken for gold due to its golden color and metallic shine. Gold, however, is rarely found in cubic form, despite its underlying crystal structure being cubic.

Instead, gold is usually discovered as irregular lumps threaded through quartz material. Prospectors panning for gold had to learn the distinction between gold and pyrite ("fool's gold").

Sodium and chlorine atoms alternate in rows in salt. The atomic centers within the cubic shape are depicted in this picture of the structure of a salt crystal. The atoms are large enough to contact one other due to their orbiting electrons. Salt was the first mineral to be studied and researched using X-rays, penetrating a crystal and revealing its structure.

Many minerals, including pyrite and gold, appear similar at first glance. One of the world's most abundant minerals, clear crystal quartz, may mimic another clear crystal, calcite, which is equally plentiful. However, based on how a crystal appears and feels, there are a few easy tests you can apply to identify even identical minerals differently. The qualities listed below are critical for describing minerals in the field or the lab.

The Crystal System is a system of crystals. On the planet, about 2500 different types of minerals have been identified. Although they appear to have an infinite variety of shapes, they fall into six primary shape groups of crystal systems. Minerals crystallize (or grow) from liquid solutions, starting at the center and expanding outward. The form of the mineral's molecule is at its center. Atoms fill a hexagonal structure in the mineral quartz, which combines with other hexagonal structures as it expands into a bigger, six-sided crystal. As a result, quartz belongs to the hexagonal crystal system, and masses of hexagonal crystals are common in rocks.

The Six Different Crystal Systems

Although most isometric crystals are cube-shaped, this method of mineral development also contains eight- and twelve-sided crystals.

A tetragonal crystal resembles a cube but is slightly longer in one direction than a cube (in this example, it is taller).

Classic prisms are orthorhombic crystals. In this system, somewhat flattened forms are widespread, as are other roughly cubic shapes that appear stretched or squashed.

The monoclinic crystal form is by far the most frequent. In this system, crystals frequently have two sides that develop at an angle to the others.

Triclinic crystals are not symmetrical in that none of the sides exactly mirror the others. There are no straight angles on the faces or edges.

In one way, a hexagonal or trigonal crystal is long, while in the other, it is six-sided.

Silver is frequently seen in the form of wispy forms known as dendrites.

Columnar prisms are formed by beryl, the mineral that gives us the jewels emerald and aquamarine.

Calcite cleaves along weak planes, resulting in lovely rhombic fragments.

Minerals form needles, blades, prisms, huge undefined structures, rounded kidney-shaped groups called reniforms, and feathery branchlike patterns called dendrites as their crystals join together. The mineral beryl, found in the blue-green emerald and aquamarine, has a habit of forming into massive prisms in rocks; the elemental material silver, on the other hand, grows in wispy, branchy dendrites rather than crystals.

Cleavage

A mineral fracture usually divides along cleavage planes, which are weak layers. Cleavage planes can be found between atom layers or in planes with weak atomic connections. A salt crystal comprises alternating rows of sodium and chlorine atoms that shatter neatly in flawless cleavage planes between them. Micas, platy, flaky minerals found in various rocks exhibit flawless cleavage between layers, resulting in thin sheets parallel to the mineral's base.

Fracture

Some minerals do not shatter along neat and even cleavage planes when struck powerfully. Instead, they break into tiny fragments with abrasive surfaces. Fracturing produces a variety of fracture surfaces, including uneven, conchoidal (curved), hackly (jagged), and splintery. A mineral can shatter in various ways, including cleavage planes on some edges and fractures where there are no cleavage planes.

On the other hand, some minerals simply fracture and leave curved or jagged edges. Opal is a mineral that fills gaps and voids in various rocks. The amorphous structure of opal contains up to 10% water. Opal breaks conchoidally when broken because it lacks a well-defined crystal structure. The hardness of a mineral is a measurement of how difficult it is to scratch it. Soft minerals can be scratched with your fingernail, but tougher minerals require a knife to scratch. Friedrich Mohs, a German scientist, invented a scale for comparing the hardness of rocks in the early 1800s. The soft, easily scratched talc is assigned a one on the Mohs scale (the ingredient once used in many bath powders)—diamond, the hardest-to-scratch mineral, with a hardness of ten. Many additional minerals, ranging in hardness from 2 to 9, are found between talc and diamond on the scale. A harder one can scratch a softer mineral. Talc, gypsum, calcite, fluorite, and apatite are among the minerals that may be scraped with a knife blade with a hardness of 512. Orthoclase, quartz, topaz, corundum, and diamond, ranked 6 to 10, cannot. Because no special equipment is required, the hardness test is simple to utilize in the field. Even nickels and dimes (hardness 312) can be used to determine the hardness or softness of a mineral.

Specific Gravity

Specific Gravity is a measure of how heavy something is. The specific gravity of a mineral is calculated by comparing its weight to that of the same amount of water. Quartz has a specific gravity of 2.65, which means it weighs slightly more than two and a half times as much as water. Gold (19.3) and platinum (19.3) are two heavyweight natural elements with high specific gravities (21.4). Sepiolite (2), graphite (2.1 to 2.3), a component of pencil lead, and sulfur (2 to 2.1), which can be found as yellowish crusts in caves and hot springs, are all minerals with low specific gravities.

In the Mohs scale of hardness, certain minerals are used to compare the hardness of other minerals and rocks.

Color

Because the same mineral can appear in various colors, color cannot always be relied upon to distinguish minerals. Quartz comes in various colors, including clear, pink, yellow, white, purple, and brown. Sapphire comes in multiple colors, including transparent, green, yellow, and purple. Different minerals, on the other hand, may have the same color. Gold and pyrite are both pale yellow minerals. Turquoise, a pale or vivid blue-green mineral stone commonly used in jewelry, can be confused with chrysocolla, another blue-green mineral. When identifying a mineral, color should be recognized but not depended upon.

A mineral streak is created when certain minerals are scraped on a porcelain tile and leave behind smudges of color. A streak is a better indicator of a mineral's identification than color since the streak is consistent across all minerals, regardless of their outer appearance. The color of the streak can be significantly different from the color of the mineral. The gold-colored pyrite leaves a green-black streak. Blood-red rubies have a pristine white streak running across them. Quartz leaves an uncolored streak and might be clear, white, gray, black, yellow, orange, red, pink, purple, brown, blue, or green.

The streak method is an excellent way to identify minerals. Minerals leave distinctive color smudges or streaks when brushed on a porcelain surface. Any mineral, regardless of appearance, will usually streak a specific hue.

Transparency

The transparency of a crystal can be defined as transparent, translucent, or opaque. Halite, quartz, and diamond are transparent minerals that allow light to pass through and can be peered through like windows. Many gemstones, including aquamarine, ruby, garnet, amethyst, sapphire, topaz, and emerald, are transparent minerals that allow light to flow through but cannot be seen. Even when cut very thin, opaque minerals like gold, silver, and platinum cannot be seen through.

Luster

The luster of a mineral describes how light bounces off its surface. Pearly, drab, greasy, silky, or glassy lusters are all possible (vitreous). The vitreous luster of many gems, including ruby, sapphire, beryl, amethyst, emerald, and topaz, gives them their bright shine. Glassy, smooth, or dull luster can be seen in gypsum, a common mineral used to manufacture plaster and cement.

COMMON MINERALS AND THEIR USES

Feldspar

Understanding feldspar begins with realizing that it may be found practically anywhere, in almost any rock. It comes in various chemical compositions based on the quantities of potassium, sodium, and calcium in its molecular structure, which also contains silica and aluminum. Orthoclase is the most prevalent potassium feldspar from albite, primarily sodium with a trace of calcium, to oligoclase, andesine, labradorite, and bytownite, which is mostly calcium with a trace of sodium, to anorthite, which is mostly calcium with a trace of sodium. Alkali feldspars are potassium- and sodium-rich feldspars. Feldspars are often light,

Sodium Feldspar

although they can also be dark or even blue in hue. They are triclinic and monoclinic, have complete or noticeable cleavage in two directions, and have uneven to conchoidal fracture. They're unique in that they're the building blocks of many rocks.

Agate

Agate is a translucent microcrystalline quartz variation. It was named after the Achates river (now called Cirillo), where it had been collected for over three thousand years by Theophrastus, a Greek philosopher.

Throughout North America, you can find river gravels, plains, mountains, and deserts along ocean and lake shorelines, as well as river gravels, plains, mountains, and deserts.

As an alluvial deposit, agates are typically found in the gravel of shorelines, riverbanks, and creek banks. Agate can also be found in igneous rock cavities. The sun makes agates glow, and rain makes them glitter as though polished, especially in locations with a lot of gravel.

Tumblers made of agates are extremely popular. Banded specimens are frequently sliced to expose the banding on a polished, flat surface. Agate slabs can be polished and displayed or used as cabochons.

Albite

The plagioclase-feldspar sequence concludes with albite. Its name comes from the Latin word albus, which means "white." it's commonly utilized as a binding substance and a glaze in glass and ceramics.

Pegmatites, felsic and igneous rocks, and low-grade metamorphic rocks all contain albite. Chemical deposition in sedimentary conditions can also cause it to form. The white tone and crystal habit of albite crystals make them identifiable. Biotite, hornblende, muscovite, orthoclase, and quartz are common companion minerals.

The majority of albite specimens are simply cleaned, clipped, and displayed. By immersing stained specimens in oxalic acid, they can be cleansed. Cabochons can be made from solid crystals. Faceting can be done on transparent specimens.

Benitoite

Benitoite, a rare barium titanium cyclosilicate, was discovered in 1907 on San Benito mountain in southern California. The rich blue tint led to the assumption that it was sapphire. California is the only place where gem-quality specimens can be found.

The most notable benitoite deposit is in the southern California mountains; modest amounts of benitoite can also be found in the Arkansas and Montana mountains.

The other few known places are claimed, and Benitoite is privately mined as a rare and precious mineral. Local rock clubs are occasionally permitted into mines to rummage through tailings. The miners will sometimes offer unprocessed gem-bearing ore in mail-order bags. Crystal specimens of benitoite are commonly obtained. This unusual material, which has a better light dispersion than diamond, forms a beautiful, faceted jewel. Cabochon cutting is an option for lower-grade specimens. It's a pretty tough stone that can be used in jewelry.

Biotite

Biotite is a family of phyllosilicate minerals in the mica group named after French physicist Jean-Baptiste Biot. Because they resemble the pages of a thick book, masses of platy crystals of mica are referred to as books. Biotite aids scientists in determining the minimum age of rocks.

Biotite is a common mineral found in igneous rock. The black masses of platy crystals make it easy to identify. Pegmatite veins, which also contain gems like beryl, topaz, and tourmaline, are indicated by biotite and muscovite mica.

Water and liquid cleaners should not be used on biotite because they can soak into the mineral and cause it to break down. The ideal tool for cleaning mica is a dry electric toothbrush. Because biotite is difficult to cut or polish, specimens are usually used for the show.

Cerussite

Cerussite is a lead carbonate that is one of the most valuable lead ores. The name comes from the Latin word cerussa, which means "white lead." it was once used to manufacture lead paint and cosmetics.

Cerussite is a mineral found in the oxidized zones of base metal deposits, especially in lead-silver deposits. Crystals should be extracted with caution, not only because they are delicate but also because they contain lead. When mining this material, gloves should be used.

Cerussite crystals are frequently cleaned, prepared, and displayed. It is feasible to facet this material; however, cutting it without breaking it is challenging. Because it is incredibly delicate, it is not suggested for use in jewelry.

Chabazite

Chabazite belongs to the vast zeolite family, and its structure and habit are similar to that of its near relative, gmelinite. Its name comes from the Greek word chabazios, which means "hailstone." The name comes from Orpheus' lyric "Lithica," which is about the mystical abilities of minerals.

Basalt cavities, veins, and vugs are the most common places to find chabazite. The crystal habit and brown-orange color tones help identify it in the field. Extreme caution should be exercised when collecting crystal clusters from the host rock to avoid damaging them. Chisels and hammers come very handy.

Although fine colorless crystals can be faceted, they are too delicate to be used in jewelry. Light soap, water, and a gentle toothbrush are generally enough to clean crystals. For a more aesthetically pleasing display, the matrix can be reduced.

Gypsum

Gypsum crystals are soft (hardness 2) and glossy and can be found near the borders of hot springs, where it is deposited from evaporating water or throughout clay layers. It might be clear, white, gray, greenish, yellowish, brownish, or reddish. White gypsum streaks It can form translucent diamond-shaped crystals, rosette-shaped clusters of crystals called gypsum roses, or tabular bundles of fibrous crystals. You can scrape gypsum with your fingernail. Gypsum is used in the construction of plaster and wallboard.

Diamond

The hardest known natural mineral generated on the earth is diamond, which is pure crystalline carbon. Its name comes from the Greek word Adamas, which means "invincible," referring to the mineral's hardness. Low-quality diamonds are employed as an industrial abrasive, while high-quality diamonds are immensely popular gemstones for jewelry.

Only the tundra of northern Canada, the mountains of Colorado, and the hills of Arkansas, where the crater of diamonds state park is located, have diamond concentrations.

Diamond is a hefty mineral that, like gold, prefers to sink into nooks and crannies. There are several accounts of gold prospectors discovering diamonds while panning for gold. Rough diamonds are also extremely smooth and do not stay pinched between fingers for long periods. Diamonds are a common faceting material. Low-grade crystals are better left as specimens for display, while high-grade diamonds are submitted to expert cutters.

Diopside

Diopside is a pyroxene mineral that forms rocks. It gets its name from the Greek word di, which means "two," and opsis, which means "vision." This refers to the two ways the vertical prism can be oriented. Chrome diopside is bright green diopside that has been dyed with chromium. The blue variety, tinted by manganese, is known as violane.

Olivine-rich basalts, andesites, kimberlite, crystalline limestones, and dolomites contain diopside. The brilliant green or blue tones of chrome diopside make it easy to spot in the field. Crystal structure and a scratch test can be used to identify colorless crystals.

Fine transparent crystals are frequently faceted, and lower-grade material is cut en cabochon. They can also be used for lapidary and jewelry.

Eudialyte

Eudialyte is a relatively uncommon cyclosilicate mineral. Its name comes from the Greek word eu dialytos, which means "easily decomposable," referring to the mineral's acid solubility. Eudialyte is a minor zirconium ore primarily utilized in jewelry and collecting.

Alkaline igneous rocks produce eudialyte. The brilliant red tones and crystal habit of fine crystals distinguish them. When extracting eudialyte, extreme caution is advised because the crystals are readily broken. To carry crystals, wrap them with paper or plastic wrap.

Only experienced lapidaries should cut and polish the eudialyte since it is extremely rare and brittle. Cleaning, potentially reducing the matrix, and displaying it are all customary practices. Acids should not be used to clean the eudialyte.

Magnetite

This common black mineral, made up of iron and oxygen atoms, is extremely magnetic, attracting iron filings and moving compass needles. It has an opaque appearance and a metallic or dull gloss. A black streak is left by magnetite. Eight- and twelve-sided crystals are common in the cubic (isometric) system. A magnetite specimen with a hardness of 512 to 612 is unlikely to be scratched by a knife blade. There is no cleavage in magnetite.

Mineral clusters form rocks. When studied more closely, this drawing depicts how a gigantic cliff of granite is made up of interlocking crystals that contain the minerals quartz, mica, and feldspar.

Fluorite

Fluorite is a fluoride mineral made up of calcium and fluoride (hence its name). It has a wide range of colors and collection spots, making it highly collectible. Fluorite is mined for use in producing hydrofluoric acid, steel, and high-octane fuels in the industrial sector.

Fluorite is commonly found in felsic igneous rocks, granitic pegmatites, and limestones. The crystal habit distinguishes it in the field. It comes in a variety of colors, with the most frequent being green, purple, blue, and colorless. Because it is delicate and easily cleaves, it must be handled cautiously when retrieving specimens.

When working with this delicate material, extreme caution is required. A mild muriatic acid solution can be used to clean heavily stained crystals. Massive fluorite can be cut and fashioned into cabochons, but its softness and flawless cleavage are not suggested for use in most jewelry.

Kyanite

Kyanite is an aluminum-silicate mineral named after the Greek word kyanos, which means "blue," referring to the mineral's prevalent blue tone. Kyanite is a raw material used in ceramics, electronics, spark plugs, and abrasives in the industrial sector.

Outcrops of metamorphosed schists, gneisses, and granitic pegmatites contain kyanite. It's most frequent in cyan-blue hues, particularly in North America. When retrieving specimens, extreme caution must be exercised to avoid damaging them.

Cutting and polishing this delicate mineral should only be attempted by skilled lapidaries. Translucent specimens, on the other hand, can produce magnificent cabochons and faceted jewels. The majority of the specimens are kept in their original state.

Hematite

Hematite, a mineral of iron and oxygen, comes in various colors ranging from blood red to iron black, but it always has red streaks. From bright metallic hexagonal crystals to dull, worn masses, it has a variety of habits and lusters. When hot liquids have passed through rock, it frequently replaces other minerals. Hematite, like magnetite, has no cleavage. It features a fracture pattern that ranges from irregular to virtually conchoidal. Hematite and magnetite are both significant iron ores.

Neptunite

Neptunite is a rare and recently discovered mineral initially identified in Greenland in the early 1900s. Its name comes from the Roman god of the sea, Neptune, and refers to the mineral aegirine, named after a Norse sea god.

Neptunite is most commonly found in the hills near Mont-Saint-Hilaire, Quebec, Canada. Specimens have also been discovered in the San Benito County highlands of California.

Neptunite can be found in pegmatites and schist and serpentinite natrolite veins. Extreme caution is required when extracting exposed neptunite crystals from the host rock.

Neptunite can be tumbled and cut into cabochons, although it's extremely rare and is typically maintained as a display piece. It's usually encased in natrolite, but with an acid wash, it can be exposed or entirely freed from the natrolite. It can also be faceted for faceted gem collectors.

Galena

Galena is a common mineral that originates in fractures in the ground when hot liquids have risen to the surface. Galena is also known as lead sulfide, a lead-sulfur combination that is a vital lead resource. It has a soft (hardness 212) and thick texture (specific gravity 7.58). It possesses perfect cubic cleavage, resulting in attractive, well-formed cubes, and is classified in the cubic (isometric) crystal system. Galena has a gleaming metallic sheen to it.

As Democritus and Aristotle suspected, all stuff comprises atoms and molecules. Atoms and molecules have crystallized into formations in minerals, which are the building blocks of all rocks. Clusters of minerals are brought together under various temperature and pressure circumstances.

Obsidian

Obsidian is a natural volcanic glass generated when extremely siliceous magma cools rapidly. Its most frequent hue is black, but it can take on other colors and have an optical appearance called shine when added minerals are added. It was named after obsius, a roman explorer who is said to have discovered obsidian in Ethiopia for the first time.

When harvesting obsidian, use gloves and eye protection because it is incredibly sharp. Obsidian's glassy surface, generally black hue, and acute conchoidal fracture make it easy to recognize. Remove chunks from a deposit or search for them as a float.

Tumbling, carving, knapping, cutting cabochons, and faceting can all be done with obsidian. It's a common material for making arrowheads out of. In a tumbler, obsidian chips easily; it is recommended that you use a buffing medium with your tumbler load.

Micas

Mica micas have a perfect cleavage that runs parallel to the crystalline structure's base. Micas grow in large clusters in rocks that layers like pages can peel off in a book. Micas have a clear streak, although the rocks themselves range in color from clear to white or gray to reddish-brown and green, with metallic shine on occasion. Micas are typically soft, with hardness ranging from 2 to 4, and may be scratched with a knife blade in most circumstances (hardness 512).

Orpiment

Orpiment is an arsenic sulfide mined commercially for use in fireworks, semiconductors, photoconductors, pigments, linoleum, and infrared-transmitting glass. Its name comes from the Latin words aurum, which means "gold," and pigmentum, which means "paint," referring to the mineral's use as a golden pigment.

Orpiment is found in low-temperature hydrothermal veins, near hot springs, and as an arsenic-bearing mineral's

alteration result. It is distinguished by its foliated masses and vivid yellow to orange tones. Because orpiment contains arsenic, caution should be exercised when gathering it. When gathering orpiment, wear a filtered face mask and gloves.

Typically, orpiment is gathered as a mineral specimen. Because orpiment contains arsenic, it should be handled cautiously while cutting, grinding, or polishing. After handling, wash your hands. Please keep it away from tarnish-prone metals since the sulfur concentration might hasten the oxidation process.

Calcite

The majority of limestones and marbles are composed of calcite. It can be found as white, transparent, gray, red, brown, green, or black crystals in various rocks. It frequently fills veins and holes in the rock around it. It has a translucent to translucent appearance and is velvety (hardness 3). Scratching it with a coin (hardness 312) or knife blade (hardness 512) distinguishes it from quartz, which is tougher (hardness 7). Calcite crystals have excellent cleavage, with planes visible on their surfaces. Cement, building exteriors, marble statues, and other decorative sculpture are all made with it.

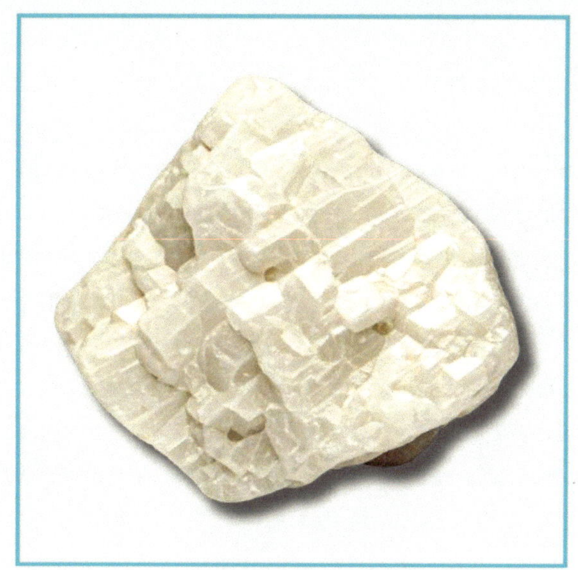

Rhodonite

Rhodonite is a manganese inosilicate that belongs to the pyroxene group of minerals. Its name comes from the Greek word rhodos, which means "rose," referring to the mineral's numerous pink colors created by manganese. It was once mined as a decorative gemstone.

Rhodonite can be found in many different manganese ore deposits. It can also be discovered as a rhodochrosite metamorphic product. Rhodonite is distinguished by its different pink tints

in the field. Rhodonite will oxidize black on the outside after being weathered out of its initial deposit. To reveal the pink interior, chip away at the edges.

Rhodonite is a versatile lapidary stone that may be tumbled, carved, face polished, slabbed, and cut into cabochons. It's tough enough to be used in most jewelry designs. Rhodonite, a rare translucent mineral, can be faceted into gemstones.

Pyrite

Pyrite is known as "fool's gold" because its brilliant yellow hue resembles gold. It can be found in various rocks as tiny square-faced crystals are strewn about. When pyrite is struck by metal, it produces sparks. On the surfaces of its crystal faces, striations, or parallel patterns of grooves or scratches, are common. Pyrite occurs in various forms, although it is most commonly seen as cubic crystals with striated (grooved) surfaces.

Thundereggs

Thundereggs, also known as lithophysae, are rhyolite-coated nodules. They're commonly made of agate or jasper, although they can also contain opal and other minerals. Their name comes from Native American folklore about two rival thunder spirits who used these mineral oddities as projectiles from thunderbird nests.

Thundereggs can be found in huge amounts of high-silica rhyolite. Their nodular shape and typically bubbly rhyolite exteriors help identify them in the field. Suspect material can be cracked open; instead of a hammer, use a rock saw to break open specimens.

Thundereggs are frequently sliced in half, polished on both sides, and presented in pairs. The agate, jasper, and/or opal inside is occasionally slabbed and used to cut cabochons for jewelry. Thunderegg fragments can be tumble-polished.

Pyroxene

Pyroxene, like amphibole, comes in various forms—diopside, augite, and hypersthene, to name a few—each characterized by molecular composition differences. Pyroxenes are usually dark in color; however, they can also be white or gray, like amphiboles. Pyroxenes are common and essential minerals in many rocks, but they require some study with a good mineral guidebook to begin understanding them. Pyroxenes exhibit strong cleavage in two directions, a hardness of 5 to 612 and an irregular fracture pattern. Pyroxenes range in transparency from translucent to opaque.

Vanadinite

A member of the apatite group, vanadinite is a hexagonal mineral. It is a valuable vanadium ore after which the mineral is named. Vanadium is mainly used to make steel more durable.

Vanadinite is a secondary mineral found in lead ore deposits' oxidization zones. Flat, brilliant red hexagonal crystals distinguish it in the field. Extreme caution is advised when extracting delicate, brittle specimens from the host rock. Vanadinite contains lead and should be handled with caution

For most lapidary applications, vanadinite is excessively soft and brittle. Mineral specimens are frequently made up of crystals and crystal clusters. Crystals can discolor and lose transparency when exposed to light for lengthy periods; therefore, keep them in a dark or lightproof container.

Amphibole

Amphibole crystals are long, prismatic, with a hardness of around 5 or 6. They are commonly found in rocks as long, prismatic crystals with a hardness of about 5 or 6. They might be black or white to gray. Amphiboles come in various chemical forms, including hornblende, anthophyllite, glaucophane, and grunerite, and they can be found in different rocks. Amphiboles feature striated prismatic to fibrous crystals with a vitreous or silky luster and good to perfect cleavage in two directions.

Variscite

Variscite is a mineral of aluminum phosphate mostly mined for collectors and lapidaries. It was named after the former German district of Variscia (now known as Vogtland), where the mineral was discovered for the first time.

Variscite is found in hydrothermal replacement deposits and brecciated sandstones as a secondary mineral. The green tones and crystal habit help to identify it in the wild. It comes in the form of nodules and venous material. Because variscite is a soft mineral, using hard-rock tools to extract it from the host rock should be done with caution.

Variscite polishes well and is a popular jewelry material despite its softness. It can be sculpted and cut into cabochons. Cutting nodules and vein material in half, polishing it, and displaying it is a common practice.

Quartz

Quartz is one of our most common minerals, and it can be found in all types of rocks. Silicon dioxide, or silica, is a chemical compound of nearly pure silicon and oxygen. Quartz can be seen as semiprecious gemstones or tiny crystals deposited in light-colored veins that cut through rocks. It's usually transparent or white, although it can be virtually any hue, and six-sided crystals topped by pyramids distinguish it. Striations (parallel lines of grooves), similar to

those found on pyrite faces, are typical on quartz crystal faces, and the crystals are frequently deformed in appearance. With a specific gravity of 2.65, quartz is not particularly dense yet relatively hard (hardness 7). Quartz crystals are well-formed hexagonal crystals that are transparent or white and cannot be damaged with a knife.

Quartzite

Quartzite is a metamorphosed sandstone with a high quartz content. Its fine-grained color and texture are generally consistent, incredibly robust, and durable. It gets its name from the German word quarz.

The large quartz grains that make up quartzite make it easy to spot in the field. It is most often tan but can be found in almost any color.

Quartzite can be tumbled, carved, and even cut en cabochon for use in jewelry. Countertops are made of large, polished pieces.

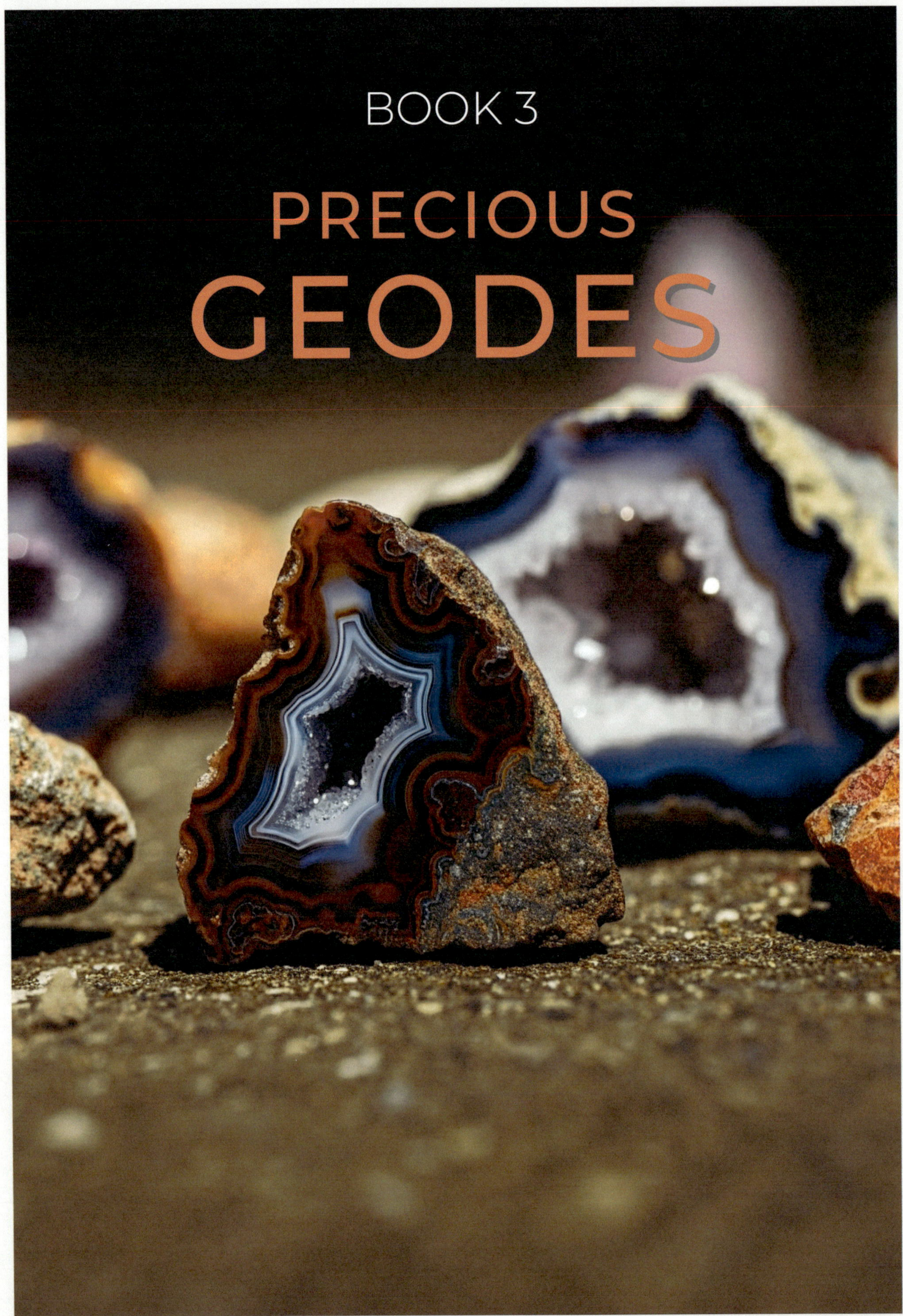

BOOK 3
PRECIOUS
GEODES

What are Geodes?

Geodes form when a gas or liquid is allowed to crystallize in a small cavity within rocks and stones. Agate is a standard crystal found inside geodes, with at least one or two layers of agate serving as a crust on most of the world's geodes. In some cases, the cavity inside a geode will be filled with crystallized chalcedony, resulting in a solid piece of agate; however, several layers of agate can coexist with another type of crystal, usually quartz growing in the geode's very center.

It might help if you knew what they are if you want to collect them, right? A geode is the Tootsie Roll pop of the Rockhound world, with a fantastic center that you can't wait to crack open. Oh, and a Rockhound is a phrase that describes people who hunt for valuable rocks and minerals, such as myself.

What exactly are they? They're rocks with a hollow center lined with gorgeous minerals or crystals that appear plain on the exterior. Igneous and sedimentary rocks are the most common places where they originate. Then some geodes are filled with crystals and are not hollow. Thunder Eggs are nodules, not geodes. However, those are still frequently referred to as the latter.

Suppose you want to understand more about this question. In that case, there are many more scientific explanations available, which you can read about later. You must know that geodes are seemingly ordinary rocks, usually round, with beautiful crystals and other minerals inside! It's exciting to collect them. They're lovely to look at and finding them is like striking gold!

Geodes can be found in a variety of locations around the world. If you reside near deserts, volcanic ash deposits, lakes, rivers, streams, or creeks, you are already sitting near some geodes just waiting to be discovered.

Make a list of suggestions before you go on your adventure. All this information can be daunting for a newbie, so having a few pointers with you when looking for geodes will be beneficial. I suggest writing down some of these suggestions and carrying them with you, so you don't forget them.

Make sure you're dressed comfortably and don't mind getting messy. Comfortable shoes are a must, and sunglasses or sunscreen are always on my list if I'm out in the sun all day.

Gather whatever items you intend to use. Although it isn't required, having a few tools on hand is usually beneficial! I always bring a hammer, a bag to put my rocks in, a couple of hand towels to dry them, and my camera since I always find interesting stuff to photograph while I'm out!

Choose a destination depending on your location. I usually go to a lake or river and walk along the banks, looking for rocks. It would help if you looked for a site with many igneous or sedimentary rocks (where pretty much all geodes will be found). Consider limestone, lava rocks, or sandstone, which are excellent geode hosts.

Notify someone of your plans, bring a companion, and don't forget to bring your phone. Even though geode hunting is a very risk-free activity, accidents do happen, and you'll want others to know where you're going just in case.

Choose the finest time of day; I like to go early in the morning before it gets too hot. Alternatively, I prefer to travel when the weather is pleasant, although this is a personal preference, and "pleasant" is subjective.

You'll need to know what to look for once you're ready. Geodes are often spherical or egg-shaped. The exterior often resembles cauliflower or has a thick appearance. In my photo, you can see what an uncut geode might look like. This one was lying along the side of Lake Barkley, just waiting to be discovered!

Check to see how light it is if you think you've located one. Geodes are typically light in weight because of their hollow interiors.

It could be a geode if it's circular, light, and has a geode-like appearance. So, here's my most valuable and favorite trick: It should be shaken. Yes, hold it up to your ear and shake it. Is there anything you can hear? Many geodes include loose crystals and sound like salt is being flung around. You may not always hear something, but I do most of the time.

So, you believe you've discovered one? The next step is to bust it open; keep reading to learn how to do it yourself!

Sock Method

This is a fantastic approach for breaking your own geode and is also one of the most kid-friendly options. You put the geode in a sock and then crack it with a hammer! It's as simple as pie, and the best thing is that all the shards and bits fall into the sock, making cleanup a breeze. Furthermore, you won't have to worry about flying shards or someone getting wounded in the process if you do it this way.

Hammering and Picking Method

Place your pick in the center of the geode, then slowly hammer the pick in a circle around the middle for a more even cut.

"I Can't Wait" Method

This is when you find a geode and break it with a larger, heavier rock.

So, pick a huge rock and crush it! You don't want to bash it too hard, but I find that a decent tap with a larger rock works simply fine. If this is your first geode and it took you a long time to find it, you might want to use the hammer and pick method to avoid smashing it into a million pieces.

COMMON GEODES AND THEIR USES

Chert

This firm, long-lasting rock is primarily quartz and comes in various colors. Many stone age tribes knapped it into tools and weapons because of its fine grain and characteristic conchoidal fracture.

Due to its endurance, Chert frequently occurs as larger rock fragments among sand and gravel surfaces. Chert is more likely to be found on the desert floor in larger rocks.

Water and a stiff brush are all that is needed to clean chert. It's an excellent rock for tumbling.

Coal

There are several types of coal, but most people think of anthracite, which is mined for heating and coal-fired power plants. The coal family members with the highest carbon content are anthracite and lignite.

Wear gloves and clothes you don't mind soiled because coal is one of the messiest rocks a rockhound can find. It's easy to spot because of its dark, almost metallic brilliance. It will also be significantly lighter than most other rocks.

Apart from carving, anthracite is unsuitable for most lapidary projects. The majority of the items gathered are simply displayed.

Geode

Geodes are circular rocks with a hollow cavity that contains one or more crystallized mineral kinds. Colorless quartz crystals make up the majority of geodes, but they can also include amethyst, citrine, agate, and other gemstones. Their name comes from the Greek word geds, which means "earthlike."

Geodes are frequently much lighter than other rocks of similar size due to their hollow chambers. Some geodes can be filled with crystals, making identification difficult without shattering or sawing them apart.

With a hammer, geodes can be cracked open. Chisels can aid in the splitting of the geode into two halves. Geodes can also be cut using a rock saw and polished on the cut faces. Cabochons can be made from flat pieces with tiny crystals.

Granite

In the continental crust, granite is the most frequent intrusive igneous rock. Pegmatites, which can contain gemstones and uncommon minerals, are found in granite. Granite is a common building material that is mined in large quantities.

Wherever granite bedrock is exposed, loose, fractured fragments of granite can be found.

Water and a stiff brush can be used to clean granite specimens.

Jet

Jet is a type of lignite coal with a jet black color and a lightweight. It gets its name from the French word for the same substance, jaiet. Ancient Rome and the Victorian era saw a surge in the popularity of jet jewelry.

There are no known deposits in Canada. Colorado, Maryland, New Mexico, and Utah are on plains along rivers and shorelines.

Jet is most commonly found in gravels and sediments, where its rich black color distinguishes it. It is the first substance to separate when gravels are washed because of its lightness.

Only use soap, water, and a gentle brush to clean the jet, then polish with a jeweler's polishing cloth. Avoid coming into contact with any hard surfaces. Jet is easily carved into beads and other ornamental objects and takes a fine polish.

Lazuli lapis

Lapis lazuli has been mined for almost 5,000 years. Commercial mining takes place in Afghanistan and Chile; however, there are a few resources in North America. Lapis is a mineral-rich rock made up of lazurite, calcite, and pyrite. Its name comes from the Latin lapis, which means "stone," and the Persian lazhward, which means "blue."

The deep to vivid blue tones of lapis, speckled with white and brassy flecks, make it easy to spot in the field. Lapis form a huge mass, necessitating hard-rock mining tools to extract specimens appropriate for lapidary applications.

Lapis polishes beautifully and can be tumbled, cut, and formed into cabochons, which are widely used in jewelry.

Marble

Marble is a metamorphic rock made out of limestone or dolomite that has been changed. Its name comes from the Greek words mármaron, which means "crystalline rock," and mármaros, which means "shining stone." Marble has been used to carve statues and monuments, as well as to decorate buildings, since ancient times.

As a talus material, marble can be collected. Look for items that can be used in lapidary projects. Hard-rock mining equipment will be required to recover huge pieces from the marble deposit. Make sure there are no visible cracks in the material you're collecting. Although the rock can be carved, tumbled, and even cut into cabochons, it is too delicate for most jewelry uses. Marble can be engraved for both decorative and practical reasons.

Limestone

Aragonite, dolomite, siderite, quartz, and pyrite are examples of minerals. Invertebrates from the sea and freshwater colors: primarily gray, with off-white, brown, tan, and pinkish tones threw in for good measure. Fine-to-medium grained, angular, rounded texture

Calcite is the main component of limestone. It is commonly formed in warm, shallow oceans when calcium carbonate precipitates from the water or when calcareous shells and skeletons of marine animals accumulate. Commercially mined limestone is used in construction materials, cement, glassmaking, and agriculture.

Unless fossil fossils can be seen with the naked eye, limestone is of limited value in lapidary. Although it is not advised for most jewelry, specimens with fascinating fossils can be sculpted and cut en cabochon.

BOOK 4
PRECIOUS
FOSSILS

PRECIOUS FOSSILS

A fossil is any preserved remains, impression, or trace of an extinct organism. Fossils are classified as minerals.

Not everyone can go fossil hunting. It's for folks who enjoy being outside and don't mind getting their hands filthy or getting a slight scratch now and again. Fossil collecting is distinct in that you can buy a lot of nice specimens for a small amount of money and build a nice collection without ever stepping out into the field. At the very least, if your budget permits it. But that's a different story; we're here to talk about the genuine thing.

So, what's the point of going fossil hunting? First, you go outside and take in the scenery. You also get to do some workouts. After a successful day of fossil hunting, you may notice certain muscles you were unaware existed in your body aching.

Second, it keeps your mind occupied. If you truly get into it, you'll learn a lot about rocks and fossils, the geography around you, and the creatures and plants that still exist on the globe. You're probably unaware of the terms brachiopod and crinoid. Don't worry; you'll figure it out like a fossil collector after a year. Your neighbors and relatives will be astounded by your extraordinary knowledge of bizarre creatures they never imagined existed in ancient times. Many of them, such as crinoids and brachiopods, are alive and well today.

Third, you will always have the opportunity to discover something truly exceptional. Either a precious fossil — some have been sold for millions! - or a fossil that has never been discovered before. Maybe you'll get a new species named after you in the end.

Do you believe it's impossible? Well, I assure you it's not. Hundreds of billions of fossils are still to be unearthed worldwide, and none of them have been investigated by scientists. And just a few thousand scientists are paid to do research on them around the world. Amateur collectors, not experts, have made many of the most important and exciting fossil discoveries ever. It's like an actual treasure hunt for adults - and, of course, children. However, it is first and foremost pure enjoyment. That, I suppose, is what a hobby should be about.

Private property includes active quarries. That is something you, as a decent person, must respect. Would you like someone to appear in your garden and dig around randomly? You'd probably freak out.

Many quarry owners are unconcerned by collectors, and others are even nice to them. Some quarries exist solely to mine fossils. These are the most effective. Expect to pay a small fee, but you can spend the entire day looking for fossils. Expect a lot of competition, and for all the excellent finds to be long gone or to go to the man next to you who just hit the ammonite jackpot as you sit there with three crappy fragments after four hours of work.

Obtain authorization from the quarry owner in any case. They usually don't like people stumbling around in the middle of their dangerous and massive machinery. So, you'll likely only have access on weekends, which is OK because I presume you're a decent person who has to work during the week to make ends meet.

So, keep the following in mind once you've arrived at the quarry. Disregard these rules and expect to be tossed out every time, without exception.

- Wear a helmet and a security vest or other brightly colored clothing. Safety goggles are recommended.
- Never, ever, ever dig a large hole in the quarry floor. Never. The guys want to move their massive tanks about, but they don't want it to resemble the Battle of Verdun.
- Leave any rock blocks that have previously been quarried and tacked somewhere for the next delivery alone. Do not touch them, even if the world's most giant ammonite protrudes from it. Of course, you can speak with the quarry owner. Maybe you can get this block for a few bucks, or he'll let you take out the fossil if he likes them, but never spoil the fruits of other people's hard work!
- Keep an eye on the quarry's expressions. Rocks fall or are blown over the edges, especially during rainy and windy weather. When there is blasting in the quarry, the rock becomes unstable. In general, it's best to stay away. There should be enough rocks in every quarry that you don't have to work directly on the quarry face.
- Leave the machines and equipment alone, of course, even though it might be alluring to drive around in that massive excavator.

Otherwise, follow the specific laws and regulations that each quarry provides. The owner is the absolute monarch and king of his quarry. Imagine yourself as a foreign diplomat representing the Fossil Hunter Nation during a period of political turmoil. You'll be treated with careful compassion and distant civility, but your every move will be scrutinized.

Take care of yourself. By doing so, you'll be able to spend many years happily collecting in your favorite quarries. As a result, you should aim to be a good, gleaming, and delicious apple; someone with whom the quarry owner might be willing to share a few beers after a few years.

How to Collect Along the Riverbank and Coast

Some countries are fortunate. One of them is England. Many English people, especially those aware of current political happenings, may disagree with this. We have a sliver of Upper Cretaceous chalk along the Baltic Sea coast in Germany, and that's about it. Everything is available in England. If Wales and Scotland are included, highly fossiliferous rocks practically cover Earth's history.

Many coastal exposures, unfortunately, have been over-collected. In the past, selfish collectors and fossil merchants have physically destroyed some, putting coastal stability in jeopardy. Because, to me, it's not about the dreaded fossils; it's about safeguarding the coast and the actual ecology and animals that exist there.

So, you can always pick up fossils at the beach in England, but you should probably leave your hammer at home. The British don't like it when their coast is sledgehammered and with good reason. You'll discover plenty of things anyway, especially if you go in the spring after the winter storms. Of course, you'll face competition from dozens of local collectors and fossil merchants, but there should be enough for everyone.

First, there are the cliffs. The cliffs are as unstable as they are in an active quarry, if not more so. The same can be said about the sides of large river ravines, lakes, and brooks. Keep your distance. Look for items at the cliff's base; there should be lots. Keep an eye out, and your ears peeled. Run if you hear the faintest hint of cracking noise. You must flee for your life. Literally.

The weather forecast is the second thing to keep an eye on. You'll have to stay at home if it's raining or storming. Let the ammonites and ichthyosaurs disintegrate – it's better than you unraveling.

Weather cannot be predicted with 100% accuracy. The tides, on the other hand – yup, the rules of physics again – can help. Every day, keep an eye on the tides and study the tidal plan. More than one obnoxious and selfish collector met his demise due to his failure to do so.

In certain regions, the tides come in quickly. They're also dangerous, though this is less of an issue along calmer coasts, such as those of the Mediterranean or the Baltic Sea. Still, what I stated about cliffs and weather holds true here.

There's one more thing. Keep your distance if a sign indicates that this is a nature protection area where a rare species of gull is lying on its eggs all day. You have no right to be there. Of course, you don't want to harm the wildlife or fragile coastal habitat excessively.

Some items will be nicely weathered if you collect them along the seaside or on a riverbed. Erosion and tidal action will almost wholly destroy a lot. Expect to locate a limited number of perfect examples in the near future. The shoreline is cruel to fossils as well.

Bringing a sieve with a mesh width of 5 mm, for example, could be helpful. You may readily sieve through soft mud and sand that smells like it came from the remains of fossiliferous rock layers. You have enough water to complete the task. There are sites where you can find a lot of lovely fossils just by doing this.

And, while you're engrossed in your collector's fever, don't forget one thing. When you're gathering in such an area, take in the beauty of the surroundings and the wonder of nature.

Dealers in Fossils

Like many fossil collectors, you may have been looking for a specific fossil for years and have had no luck finding it. Eventually, you may decide to descend into the depths and attend a fossil show. This way, you can get the fossil you desperately need from one of those "shady dealers" everyone talks about with their hands over their mouths—paleontology's boogeymen

Fossil dealers are seen as outcasts, and many paleontologists, both professional and amateur, hold them in low regard. They are dreaded because they violate the law. Rumors say that many have spent time in prison at least once in their lives, and some have even murdered to obtain a million-dollar specimen.

But should you believe in hearsay?

Most fossil dealers are brilliant individuals with superior geological and paleontological expertise that would put most university professors to shame. They have firsthand experience gained via hard labor.

They could have all been sitting on their bottoms in some bureau, day in and day out, waiting for their paychecks and never having worked a day in their lives. Instead, they go out to the fields, moving tons of rocks and working with dangerous machinery and equipment. They unearth massive fossils while getting soiled daily, becoming filthy, and sustaining injuries.

They do all this despite the risk of going into prison, attempting to avoid the snares and traps of the ever-increasing – and mostly illogical – "fossil protection regulations."

Though often seen in a bad light, fossil dealers are people running a business to earn money for whatever reasons they may have – be it to simply get rich or to purchase food and clothing for their families.

So, a few questions may pop up in your head. Is it legal to purchase fossils? Is this a good method for displaying a collection?

The answers are "Absolutely" and "Yes, if you're a millionaire."

Expensive specimens are sold by fossil dealers, at least the big ones. For a few bucks, you're unlikely to locate the one small ammonite you've been missing from your favorite spot. However, now and then, you will come across something that appeals to you as lovely and unusual. From a faraway part of the globe that you will almost certainly never visit.

Fossil sellers are in business to make money. They work in the business world. They want to acquire as much money as possible for their belongings. They are in charge of restoring specimens and specialize in composite specimens. They change the color of the specimens and may sometimes fabricate them because they are compelled to do so. Due to current Chinese legislation, it is presently impossible for local fossil merchants to export real dinosaur eggs without incurring jail time.

If you want to buy fossils, the more information you have, the greater your chances of getting a good deal. The vast majority of fossil sellers resemble vendors in an Arabian bazaar. They enjoy bargaining with a good customer. The less you know, the more likely you are to be duped, and that the lovely all-spiny Moroccan trilobite you got for a few hundred dollars would dissolve almost totally when you put it on the radiator for too long in the winter.

Do not purchase if you are doubtful. If you know many seasoned collectors, ask them for recommendations on which merchants to trust. Many fossil displays, at least in Germany, have specialists from local museums and universities on hand to answer questions about whether or not a specimen is genuine. In Germany, fossil merchants are not held in such low regard as in many other nations. This could be because some of the best German paleontologists were fossil merchants at one time or another. That has a lot to do with the country's tumultuous history.

Fossil merchants have existed for as long as paleontology has existed as a science. They've also crammed the major museums. What would the major American museums be like if they didn't have a Sternberg? What would the marine reptile gallery at the British Museum be like without Hawkins? Without Hauff, what would the Stuttgart State Museum look like?

History of Fossils

People have studied footprints, bones, and shells in rocks for thousands of years, labeling them according to myths and fairy tales. Elephant tusks were once referred to as unicorn horns. Toadstones, or worn animal teeth, were thought to be derived from dying toads and were said to cure certain diseases. The oyster Gryphaea's thick, curved shells were dubbed "devil's toenails." Circular seashells were considered coiled snakes turned to stone by an enchantress. None of this speculation was later proven correct; the fossil hunters were only correct because their finds were the remains of creatures from the past.

A plant or animal must be buried by sand, silt, mud, or other sediments in a river, lake, pond, marsh, or other habitat before it can become a fossil. Although many organisms die and decompose on the ground before burial, others are protected from rotting by being covered first. Burials are likely underwater, where currents can rapidly move sediment over dead creatures. Windblown sand, volcanic ash, river mud, and natural tar can cover land organisms. If a plant or animal is buried soon after death, its bony or wooden hard parts—bones, shells, teeth, and branches—may be preserved in the enclosing sediments.

Fossil mammoth tusks were once thought to be unicorn horns, and Jurassic-period oyster fossils were supposed to be devil's toenails.

Fossil ammonites, such as this Jurassic-era Hildoceras, were considered coiled snakes turned to stone by an enchantress. To complete the image, artisans carved snake heads on the ammonites.

Toadstones, which were thought to be coughed up by toads, are the fossils of the powerful crushing teeth of the heavily scaled Jurassic-age fish, Lepidotus.

Body parts can be preserved in a variety of ways. Burial may be quick enough, or the organism's environment may be so hot and dry that a body is preserved nearly intact, both chemically and structurally. Or the hard parts may petrify (turn to stone) when rainwater seeps into microscopic pores in the bone or wood, leaching away the original material and filling the tiny holes with harder minerals. Fossils are also preserved when a creature's body parts dissolve completely after burial, leaving behind an empty mold in the rock. A fossil cast is formed if the mold later fills with mineral matter carried by seeping rainwater.

Death, burial, and discovery: A fish dies and drifts to the seafloor, quickly covered by mud and silt carried by ocean currents. Its body decomposes, leaving only the hard bones, gradually replaced by harder minerals carried by water seeping through the sediment. More layers of sediment bury the fossil skeleton even deeper. Over millions of years, the earth's surface changes, and the sea ebbs, leaving the rocks on dry land. Millions of years of erosion may have stripped back the rocks, exposing the bones well enough for humans to find and excavate.

Petrified remains of plants and animals are common in the fossil record, but other types of fossils are also significant. Burrows, tracks, footprints, eggs and shells, nests, and droppings are examples of trace fossils.

An organism may leave only a trace of its existence. Trace fossils are preserved signs or trails, such as ancient footprints, gouges left by dragging animal tails, worm burrows, egg and shell remains, animal nests, and animal droppings. Or the traces could be from human activity, such as ancient stone tools or weapons, which are sometimes discovered alongside animal fossils. Inclusion fossils are objects trapped in ancient evergreen trees' hardened sap or amber. Insects, spiders, small lizards, and plant fragments preserved in perfect detail could be included.

The history of fossil study is long and varied. Although speculation about fossils dates back thousands of years to the inquisitive Greeks, modern paleontology—the study of ancient life—began closer to the 1600s. In 1667, Nicholas Steno (the Danish physician from Chapter 4) became fascinated with small, sharp stones known as tongue stones in larger rock beds on the Mediterranean island of Malta. Steno used his knowledge of anatomy to deduce that the tongue stones were ancient shark teeth. He proposed that Malta was once underwater and that the rocks containing the shark's teeth had been deposited by ocean water over a long period of time. As a result, he was the first known scientist to document fossils as the remains of long-dead creatures—a fundamental principle of paleontological study. Steno's laws, the law of superposition and the law of original horizontality, were derived from observations made during his numerous geologic excursions and are critical to understanding rock deposition.

William Smith, a pioneer in British geologic discovery, discovered that rock layers, or strata, can be identified by the fossils they contain.

An engineer and surveyor, Smith created the first useful geologic maps from his journal notes on fossils and strata found throughout England.

Following Steno, scientists added observations to the growing science of paleontology. While examining rock layers for a coal-mining project in the early 1800s, British land surveyor William Smith noticed identical groups of fossils in similar rocks across England. He illustrated the fossils and their surrounding rock units, concluding that rocks containing similar fossils must be of the same age, wherever they are found in the world.

The work of Steno, Smith, and others of equal importance indicated that fossils must be roughly the same age as the rocks that contain them. Younger rocks, therefore, contain younger fossils than older rocks. Unless rocks have been overturned or bent by crustal folding, layers of younger rocks with younger fossils overlie older layers of both.

First Fossils

Algae collections are the oldest known fossils, dating back 3.5 billion years. Scientists believe that the sun's rays heating the earth's first oceans created ideal conditions for developing proteins and chains of complex chemicals in seawater. These simple life forms changed or evolved into more complex forms, such as single-celled, bacteria-like algae. As they died, they fell and piled up in huge mats on the ocean floors. When they fossilized, they left behind massive beds of limestone and stromatolites, which are mounds of layered blue-green algae that resemble rock cabbages. Stromatolite deposits were discovered in large quantities in Canada's Gunflint Chert, a 2-billion-year-old ocean-deposited rock.

Algae collections are the oldest known fossils, dating back 3.5 billion years. Scientists believe that the sun's rays heating the earth's first oceans created ideal conditions for developing proteins and chains of complex chemicals in seawater. These simple life forms changed or evolved into more complex forms, such as single-celled, bacteria-like algae. As they died, they fell and piled up in huge mats on the ocean floors. When they fossilized, they left behind massive beds of limestone and stromatolites, which are mounds of layered blue-green algae that resemble rock cabbages. Stromatolite deposits were discovered in large quantities in Canada's Gunflint Chert, a 2-billion-year-old ocean-deposited rock.

The planet's early atmosphere lacked oxygen, and the blue-green algae that comprised the stromatolites did not require it to survive. However, as they drew energy from the sun, they digested their food and expelled oxygen as waste. After thousands of years of countless algae continuously producing oxygen, the composition of the earth's atmosphere changed. Larger, more complex animals evolved and thrived in the presence of oxygen 1.5 billion years ago. Fossil finds in sedimentary deposits in Australia's Ediacaran Hills revealed well-preserved evidence of soft-bodied ocean animals from 670 million years ago: jellyfish, sea pens (which resemble ferns), and arthropods (ancestors of today's crabs, lobsters, and insects).

Stromatolites the size of footstools have been discovered onshore at Shark Bay in Western Australia.

Stromatolites form in the following manner: Algae grow on mud; a layer of mud or silt washes in and sticks to the algae, allowing new algae to grow on top; more layers alternate and heap up in mounds.

Backbones and Shells

The first creatures of the ancient seas had no backbones, teeth, or other easily preserved parts to contribute to the fossil record, but that changed when hard-shelled creatures entered the picture 570 million years ago. Because more hard parts were available to preserve, fossils have been abundant in rocks deposited since that time. Animals most likely developed shells due to a change in environment—perhaps the acidity of the sea changed. Brachiopods

(clam-like creatures), graptolites (wormlike animals in branched colonies), trilobites (animals with outer skeletons and jointed legs), coral, jellyfish, sea sponges, and sea anemones came to life in the oceans. Some primitive fishes developed backbones. Early fishes had bony plates covering their heads, rows of scales on their sides, and no fins or jaws, according to fossils dating back around 500 million years.

A fossil brachiopod (lamp shell) from the Pennsylvanian period is embedded in limestone. A section of the stem of a fossil crinoid, or Pennsylvanian-age "sea lily," is shown as it once lived, attached to the ocean floor by its stem.

Soft-bodied creatures survived, adapting to changes in the oceans. Although the soft-bodied survivors left few remains in the fossil record, the Burgess Shale, a significant 530-million-year-old fossil deposit in Canada, provides a glimpse of the wide variety of shapes assumed by soft-bodied creatures in those early years. Their bodies were preserved in incredible detail in the fine-grained shales, which were thought to have been deposited by ocean currents spilling over into nearby lagoons.

Between 400 and 350 million years ago, oxygen in the atmosphere, which had been gradually increasing since the beginning of life on Earth, became abundant enough to ring the planet in an ozone layer, much like today. Living things could survive without water if they were shielded from the sun by ozone. Over many millions of years, as life flourished in the oceans, some moss-like sea plants washed up on land and took root, becoming the first creatures to live on shore. Amphibians evolved from developing fishes 395 million years ago, developing lungs that allowed them to live for short periods out of water: Extensive deposits of 380-million-year-old Old Red Sandstone in Scotland and the United States (in New York, Pennsylvania, and West Virginia) contain fossils of lobe-fin fishes, creatures with both fish and amphibian characteristics—fish scales as well as strong-boned, lobe-shaped fins that later evolved into feet.

A Great Extinction

Forests of horsetails, ferns, and cone-bearing trees covered the land 350 to 250 million years ago. With their more primitive lungs, reptiles colonized areas further inland than amphibians could. Reptiles also dominated the seas and the air, evolving both swimming and flying forms. Rocks deposited in rivers, swamps, lakes, and oceans 300 million years ago are rich in fossils of ancient reptiles, fishes, woody plants, mollusks (shellfish), coral, horsetails, land snails, centipedes, millipedes, and cockroaches.

The fossil record from 225 million years ago, on the other hand, shows a significant break in evolutionary lines. More than 90% of all plant and animal species perished in an extinction even greater than that of the dinosaurs 160 million years later. Many scientists believe the 225 million-year-old mass extinction was caused by gradually drying the earth's shallow seas. The shrinking of the oceans left fewer places for plants and animals to survive. Half of all jellyfish, sponges, mollusks, worms, and fish species were wiped out. Only four of the fifteen major groups of reptiles survived. Trilobites, horned corals, and many brachiopods perished. Only about five out of every hundred species remained. All life on Earth today evolved from the few survivors of the great extinction, filling empty niches in novel ways.

Some ocean reptiles survived, relocating to land as the oceans dried out. These new land reptiles resembled modern crocodiles—large lizard-like creatures with muscular hind legs. They also had distinct hip joints that eventually allowed them to walk upright—as ancestral dinosaurs.

The Dinosaurs

In 1822, British fossil collector Mary Ann Mantell discovered a fossil bone that resembled a massive stone tooth while searching for a road under repair. Further investigation revealed associated bones resembling the reptile iguana, only sixty times the size. Scientists who studied Mantell's discovery named it Iguanodon (meaning "iguana tooth") after discovering that it no longer existed on Earth. In subsequent years, fossil hunting around the world turned up fossils of other giants and extinct reptiles with a curious difference from today's reptiles. The

extinct creatures appeared to have walked with their feet set beneath them. Reptiles nowadays walk with a wide, splayed-out stance. Scientists decided to name a new animal family based on fossil evidence. Richard Owen, an anatomist at the London Natural History Museum, proposed the name Dinosauria, which means "terrible lizards," for the ancient reptiles in 1841.

Dinosaurs achieved great diversity in their time, ruling the world for 160 million years. The fossil evidence indicates a wide range of shapes and abilities: some dinosaurs had long necks for browsing treetops like giraffes, some were meat eaters who could chase their prey at high speeds, and some had wings and thus could fly. Hundreds of dinosaurs have been discovered worldwide, and some paleontologists believe thousands more will be discovered in the future.

Despite their amazing adaptations and family success, dinosaurs vanished from the fossil record around 65 million years ago. Some scientists believe the mass extinction was caused by gradual climate change. Others believe the mass disappearance was caused by a global catastrophe, such as a meteor colliding with the Earth or widespread volcanic eruptions. Others believe that a deadly disease spread throughout the world, carried from continent to continent by the roaming dinosaurs. Because the dinosaurs died out, other species of animals were able to thrive.

Mary Ann Mantell, a fossil collector by hobby, discovered the remains of the dinosaur Iguanodon in rocks dug up during road construction.

Dinosaurs walked with a better stance than other tetrapods or four-legged animals. Lizards, for example, sprawl with their legs out to the side. Crocodiles can rise and move quickly with their bodies above the ground when necessary. But dinosaurs had pillar-like legs that supported their bodies from directly beneath, similar to horses. Dinosaurs were thus more efficient than lizards or crocodiles at running and supporting their heavy body weights.

Beyond the Dinosaurs

While dinosaurs ruled the earth, other plants and animals thrived in the land's mild, moist climates. Plants like ferns, ginkgo (maidenhair) trees, cone-bearing trees, and cycads evolved quickly (palm-like tropical plants). Flowering plants evolved toward the end of the dinosaur era, providing them and other land creatures with a rapidly reproducing and spreading food source. Ammonoids, oysters, and brachiopods evolved and spread widely in the oceans. The oceans were ruled by large, meat-eating plesiosaurs and ichthyosaurs, swimming reptiles that were not dinosaurs. Pterosaurs, winged reptiles that gradually evolved to the size of today's small airplanes, took to the skies. Other reptiles, such as crocodiles, turtles, and lizards, spread and left abundant fossils in rocks dating from 200 to 65 million years old.

The first birds evolved during the dinosaur era, as discovered in the Solnhofen limestone in Germany, a 140-million-year-old limestone fine-grained enough to preserve detailed impressions of fossil crayfish, jellyfish, pterosaurs, insects, and dinosaurs. Among the impressions of delicate features in the Solnhofen deposits are the world's oldest known fossilized feathers, associated with Archaeopteryx, a crow-sized bird with teeth and clawed wings—reptilian characteristics.

As this illustration shows, all living animals may have lobe-finned fish and early tetrapods as common ancestors. Dinosaurs and other well-known creatures diverged from early reptiles at various times in response to environmental changes. Some animal species have survived to the present day, but many have not.

One group of dinosaur-era land animals survived by remaining small, hiding from the great reptiles, eating insects, and foraging primarily at night. When the dinosaurs and giant reptiles died 65 million years ago, these small, adaptable creatures—mammals—survived alongside the smaller reptiles (crocodiles, snakes, lizards, and turtles), amphibians, bony fishes, and birds. Over 60 million years, modern life forms evolved from early mammalian rodents and primates (apes and monkeys) to large flightless birds, flowering plants, fishes, sharks, and sea clams and snails.

Fifty-million-year-old fossils reveal life forms that survived and evolved after the dinosaur extinction. Preserved horse teeth and foot bones show that early horses were short, five-toed creatures that gradually increased in size and developed the faster, tougher hoof of the modern horse over millions of years. Fossils of early barnacles, sea urchins, sea snails, and whales show that their shells and skeletons changed over time—for example, sea urchins took on many shapes, and ancient whales gradually acquired teeth. Fossils of early tree species indicate that they became less common as flowering grasses spread, displacing forests in many areas. The fossils of animals that roamed the new grasslands show large herds of grazing animals such as pigs, camels, rhinos, antelopes, and horses. Fossil apes are common in European and African rocks from 25 million years ago. And fossils of human-like primates date back 10 to 20 million years.

The Human Record

We've only known about fossil humans for about a hundred years. Only a few specimens of fossil humans had been discovered in Africa and Asia by the 1880s. Anthropologists, or scientists who study human fossils, now work at sites worldwide. Human fossil discoveries are becoming more common as searches become more intense, and each discovery raises new questions about the human family tree.

The puzzle of human evolution is incomplete because fossils are rare in general and only provide fragments of information on how we began. The oldest humanlike bones were discovered in Ethiopia, Africa, in the early 1970s, in an excavation of at least fourteen individual early humans known to scientists as The First Family. The bones of a 4-million-year-old female named Lucy scientists revealed that she walked upright, had hands and feet like humans, left footprints similar to humans, but had a brain about one-fourth the size of ours.

Scientists have discovered that Lucy's ancestors died out more than a million years ago. Human fossils from later times, the remains of a family line known as Homo (man), show a larger brain size, roughly the same as we have today. Homo erectus, our large-brained, upright-walking ancestors, could stand and search the growing grasslands for the animals they hunted. They could carry food home in their arms to growing communities of families, fashion tools, and scrape hides for clothing. They lived well enough off the meat of the fantastic beasts of their time—woolly mammoths, saber-toothed tigers, giant wolves, and sloths—to follow them on their migrations across the globe. Their intelligence, combined with the physical changes that came with an upright stance, remained as the line evolved into Homo sapiens, the species of human that survives to this day.

Some equipment is required for fossil hunting. When you go to gather fossils, you will encounter hard rocks. Some are unpredictable and can roll on their own, so to protect your toes from being crushed, consider bringing steel toe caps and ankle-protecting high boots. Rock edges are sharp and hard – enough to sever your limbs if you let them. To prevent this, you can wear regular work gloves.

- Goggles for safety.

- Bring a good rucksack with plenty of space. One that feels right for you. The best places to collect are in the middle of nowhere. Fossil collecting is not for slackers who want to park their big rig right in front of a perfectly preserved dinosaur skeleton peering out of the cliff face, ready to be collected. You'll need to walk—a lot. You'll also want to make room for the items you find and store them safely and for water and food. Even in cold weather, always bring enough water on longer trips. Work is involved in fossil hunting. You will sweat. A good rucksack is your best friend.

- Bags made of plastic. You will discover small fossils. You'll find broken fossils that will be nice specimens once glued back together. Broken fossils are not an issue. The large dinosaur skeletons displayed in museums have been pieced together from broken bits and pieces. So, if you find a nice but broken fossil, put them all in one bag, and you'll be able to put them back together later. If you throw them in your rucksack randomly, you won't be able to fix them.

- Something to tie everything together.

- Newspapers. The best newspapers are usually old ones. Rocks are rough; throwing your fossils into your rucksack without protecting them from each other's rough surfaces may ruin them completely. Wrap each item individually. Your fossils will express their gratitude by remaining beautiful.

- Keep a few boxes in your car for storing your finds on the way home. Place the large and heavy items in those boxes first, followed by the small and fragile items.

- A handheld lens. Magnification 10x. Some fossils are pretty small. They can be challenging to see when a rock partially hides them. A hand lens is beneficial.

- A chisel isn't always necessary (except when working with fine-grained, strongly layered sediments like shale or slate), but you can bring one or two.

- The most important thing is the Hammer (with a capital H). The hammer of a fossil collector is analogous to a cowboy's colt. He never leaves the house without it. He looks after it and looks after it like an old friend. And it must be a true geological hammer, not some crappy old wooden thing from grandpa's basement. Expect to spend between $50 and $100. You will receive something that will hopefully accompany you for the next twenty years, if not longer.

The distinction between a regular hammer and a geological hammer is vast. You'll never go back once you've tried one. The most important consideration, however, is security. A good geological hammer does not break easily; it is made of high-quality steel that will not splinter when you hit a tough rock with all your might to extract that beautiful oyster or whatever you have discovered. Don't hesitate to invest in your safety.

Anything else is unnecessary as a newbie until you gain experience except perhaps a first-aid kit. As a beginner, you will occasionally collide with the rocks, and they will usually win. Having a first aid kit on hand may not be a bad idea. And, of course, having company, whether family or friends, is always a plus. It's also helpful in case something goes wrong. Always take your phone or smartphone with you and keep your emergency phone numbers handy if you go alone.

Locating a Specimen

Look for cliffs or ledges when prospecting. Pegmatites are often white or light-colored rocks. Search as far off the beaten path as possible, but keep in mind that the Maine woods are vast; they all look the same, and there are many areas where cell phones do not work.

Look for large pieces of mica or feldspar along the bottoms of cliffs. Any quartz crystal you might find had to have come from a pocket. A pegmatite most likely formed any piece of glassy or clear quartz. Follow the peg minerals uphill to their origin. Evergreen trees frequently grow in areas with thin soil, indicating that the ledge is close to the surface.

Look for pegmatite minerals (mica, feldspar, and quartz) in streams and brooks and try to trace them back to their source. Pegmatite may be underwater but may extend on both sides of the brook. The degree of jaggedness indicates how close the source is — very sharp edges suggest that a glacier most likely transported the rock from nearby. The rounded edges indicate that it came from a long distance away.

COMMON FOSSILS AND THEIR USES

Ammolite

A rare gem variety of cretaceous ammonite fossil known as ammolite. The fossils are mostly aragonite but may also contain calcite, silica, pyrite, and other minerals. Iridescent colors include green, blue, indigo, violet, red, yellow, and orange.

Ammolite mining areas in Canada are found in ancient river gravels and along river banks. All known sites are privately owned or leased tribal land and are not accessible to unauthorized collectors. Gem and mineral shows and clubs can sometimes provide access to mines for rockhounds. Ammolite is distinguished by its intense display of iridescent colors.

Because ammolite is soft, specimens must be polished with care. To protect finished material from scratches and drying, it is frequently sealed with polyurethane or epoxy.

Ammonoid

Ammonoids are a type of extinct cephalopod that can only be found as fossil evidence in ancient seas. Ammonoids and dinosaurs went extinct at the end of the cretaceous period.

These fossils are distinguished by their spiral-shaped shells that are divided into chambers. Extreme caution should be exercised when extracting fossil specimens because they are easily damaged or broken. Broken specimens can be reassembled with glue. The packing material such as newspaper or bubble wrap can help you get your specimens home safely.

Many ammonite fossils can be displayed in their natural state. For a better display piece, specialty tools can help expose fossils from the host rock. Preparing fossils requires caution because they are brittle.

Belemnite

Belemnites are a type of extinct cephalopod that existed from the Jurassic to the late Cretaceous periods. They looked like modern squids but had a calcitic guard or shell that protected them from predators. The calcitic guard is the part of the belemnite that can be found as a fossil today. Belemnite was designated as Delaware's state fossil in 1996.

Belemnites can be found in marine sedimentary rocks and are distinguished by their bullet-shaped calcitic guard or shell. Suture lines, common in cephalopod fossils, can also be seen in them. Be careful when extracting belemnites from their hard host rock because they are easily broken. Broken specimens can be reassembled with glue.

If the shells are exposed sufficiently, fossils can be displayed in a natural setting. An experienced prepper can expose more of the fossil with specialty tools such as an air scribe to give specimens a more aesthetically pleasing appearance.

Inoceramus

Inoceramus is a genus of extinct bivalve mollusks known as pelecypods (clams) that thrived in ancient seas. The name is derived from the Greek word for "strong pot," alluding to the shell's strength and thickness. Inoceramus fossils are frequently adorned with gleaming mother-of-pearl coatings.

Look for soft gray shale exposures along creek and river banks. These fossils are frequently collected from the surface, but some digging may be necessary.

Water and a soft brush should be used to clean inoceramus fossils. Cleaning should be done with care because mother-of-pearl coatings, if present, can be fragile.

Crinoid

Crinoids are echinoderm marine animals that include starfish and sea urchins. They are made up of a segmented stem, branches, and leafy arms. Crinoids are commonly fossilized in limestone and shale.

The best crinoid-collecting locations are weathered limestone and shale exposures. Small fossils, such as the disklike segments of crinoid stems, can be recovered using screens.

Crinoid fossils are delicate and should only be cleaned with water and a soft brush.

The nautiloids are a diverse group of marine cephalopods. While nautiloids were once widespread, the pearly nautilus is the only genus that still exists today. The nautiloids' first shells were straight and conical, but they later developed whorled shells. Both types have been discovered as fossils.

Nautiloid fossils are easily extracted from soft, weathered limestone and shale formations. Hammers and chisels are required to remove these fossils from harder rock. It is critical to avoid damaging the fossil.

Nautiloid fossils are highly delicate and should only be cleaned with water and soft brushes. Working slowly and carefully with knife points and dental picks, bits of matrix rock can be removed.

Petoskey Stone

The Petoskey stone is a fossilized coral. It is derived from Devonian deposits of rugose coral and hexagon area percarinata. It was designated as Michigan's state stone in 1965. Pet-o-sega, the Ottawa chief, inspired the name.

Throughout North America, along lake shorelines and in nearby river gravels. The eastern shoreline of Lake Michigan is the primary collecting area; specimens can also be found in Illinois, Indiana, Iowa, and Ohio, as well as in southern Ontario, Canada.

Dry Petoskey stone can resemble regular limestone. It is easily identified when wet by its gray coloration and the six-sided pattern formed by coral growths.

Petoskey stone is an exceptionally soft stone that can be easily carved, shaped into decorative items, and cut into cabochons. This fossil can be shaped and polished using simple hacksaws and sandpaper.

Petrified Wood

Petrified wood is a fossil in which the original wood has been replaced, most often by quartz varieties such as chalcedony and jasper, but sometimes by opal. In 1975, petrified wood was designated as Washington's state gem.

Petrified wood is easily identified in the field because it resembles actual wood that has turned into a rock-hard mineral. Good examples show cellular replacement and the wood rings commonly found inside. On the outside, you'll notice bark markings and knots.

Petrified wood that has been replaced by agate, jasper, or opal polishes beautifully and has many lapidary applications, such as tumbling, carving, polishing, sphere-making, and cutting en cabochon. Large, well-prepared specimens make for eye-catching displays.

BOOK 5
TOOLS

TOOLS

One of the best aspects of this hobby is that it is entirely free! You don't need special equipment; you could spend all day collecting geodes and cracking them open. You'd need a larger rock to crack them with (though they might not break as neatly). Having just said, I do find a few tools helpful and wanted to share them with you.

Clothing and Safety Equipment

The proper clothing and safety equipment is essential for a successful and enjoyable rockhounding trip. The right rockhounding outfit can make or break your day.

Before venturing into the desert, forest, mountain, or other unpredictable terrains, make sure to put on a proper pair of kicks. It's easy to prioritize comfort over practicality, especially if your preferred rock-hounding location is a long distance away, but don't fall into this trap. On more than one rockhounding trip, I twisted an ankle by wearing casual sneakers with no grip instead of a sturdy pair of hiking boots.

The following sections cover proper rockhounding attire and safety equipment in greater depth.

When you're out in nature, the weather can change quickly. For example, the desert can be scorching hot during the day and almost freezing at night. A sudden desert storm can catch a rockhound off guard. So be ready for anything. Before you leave, check the weather forecast, but also pack or wear the following:

- A rain jacket
- A tough pair of pants keep your legs safe while digging and hard rock mining.
- A hat to protect yourself from the sun
- In case of a temperature drop, bring a warm sweatshirt or jacket.
- For potential sandstorms, wear a long-sleeved shirt and long pants.

Wear what is comfortable for you, but keep in mind the types of specimens you will be looking for and the location of your rockhounding.

Footwear, like clothing, depends on what you intend to do and where you plan to do it. A good pair of hiking boots is essential if your rockhounding trip includes some hiking.

Wear waterproof shoes when exploring wet areas such as rivers, lakes, and oceans (or on rainy days on land), especially if the water is cold. Rubber boots or waders are popular for crossing shallow creeks or rivers and preventing your feet from getting wet. In warmer waters, a pair of sturdy sneakers you don't mind getting wet or some aqua socks will suffice.

If you plan on pounding on rocks with a hammer, you should wear proper eye protection. An eye injury is the last thing you want to happen to you, especially when you're out in the field and don't have easy access to a medical facility. Find the most comfortable protective goggles or glasses that provide complete eye coverage. Regular prescription glasses or sunglasses are better than nothing but will not protect your eyes from flying rock.

When rockhounding, a thick pair of gloves can be not only helpful but also necessary. The shards of obsidian can be extremely sharp. Many other minerals may have sharp edges and can easily cut or stab your skin. Gloves also protect your fingers when moving rocks on talus or overburden hillsides. It's also a good idea to wear gloves when hammering rocks to reduce the numbness that can result from prolonged hammering.

First Aid Kit

A proper first aid kit is an essential thing to have with you when rockhounding. It can truly make or break an emergency. Scrapes are inevitable when your hobby includes digging and hammering. As a result, bandages are the most commonly used item in a rockhound's first aid kit.

- Gauze
- Trauma pads
- EMT shears
- A splint
- Alcohol swabs
- Antiseptic wipes
- Antibiotic ointment
- Safety pins
- Tweezers
- Cotton swabs
- Medical tape
- Nitrile gloves
- Any allergy medication that is required
- Itching, pain, fever, inflammation, and diarrhea medications
- When rockhounding in areas with venomous snakes, a snakebite kit is recommended.

Suppose you intend to do some serious hiking while rockhounding. Not only do you need a first aid kit but also a survival kit. Even the most experienced outdoor enthusiasts can become disoriented or injured. It is advantageous to be prepared.

When rockhounding in the field, bugs are almost unavoidable. Some people swear by DEET-containing products. Others prefer making their repellent. Find what works for you and keep it with you at all times. Being attacked by tiny things that are constantly trying to bite or sting you can ruin an otherwise enjoyable rockhounding trip.

Protection from the Sun

The majority of rockhounding is done in warm to hot weather. Sunscreen with an SPF of 30 or higher is essential for prolonged sun exposure on cool and cloudy days. Rockhounds who intend to dig in the exact location for an extended period will often erect pop-up tents or string-up tarps to protect themselves from UV rays. Don't forget your sunglasses. They not only help to block the sun's rays, but some polarized varieties can improve your vision when looking for specimens in rivers, creeks, and lakes.

Navigation Tools

It is critical not to lose sight of where we are in nature. Fortunately, we have navigation on our smartphones.

It's not enough to rely on your smartphone; you should have a device that knows where you are. It could be as follows:

- **GPS Device**: This is helpful because many rock fishing areas lack cellular or internet service, meaning you can't always rely on your smartphone to return to your car or camp.
- **Compass**: Because high-quality GPS devices can be costly, a compass may be a better place to start.
- **Topographic maps, atlases, and/or regional guides** (Guide): These additional items will assist you in navigating to and from your preferred rock location.

Don't be misled by our extensive list of accessories! Most people begin with a basic set and gradually add to it. You may only need a few tools depending on your location, the type of material required, the weather, and other factors.

A lot of planning goes into a successful rockhounding trip. From determining the best route to a site to anticipating any unexpected detours along the way, a little planning and a few useful navigational tools can ensure you arrive at your destination. It's also a good idea to keep records along the way to remember any problems you had getting to a location and what you found there. The following sections go into greater detail about the best navigation and record-keeping tools and equipment.

Pencil and Notebook

A notebook and writing materials are helpful for both planning a rockhounding trip at home and collecting in the field. Although you can make lists on your phone, experience has shown that using a phone in the field is not always easy. Your hands may be too dirty to operate the touchscreen and buttons, and bright sunlight may make it difficult, if not impossible, to see the screen.

You may come across other rockhounds in the field who can lead you to a new or better site nearby. Make a note of the instructions! Even if you think you'll remember after three turns, you might forget whether the fourth turn is a right or a left.

Compass/Altimeter

Having a compass with you when you're out in nature is always a good idea. It's easy to become disoriented while collecting minerals in the field, especially if you're exploring a large area and mostly staring at the ground. Always know which way to walk from your vehicle to a rock-hounding location (north, south, east, or west).

When your rockhounding trip takes you up a steep hillside or mountain, and you need to find a deposit at a specific altitude, altimeters come in handy. An altimeter saves you the trouble of walking too far or not far enough.

GPS

Even with the best maps and directions, locating a small deposit in the wild can be challenging. A GPS can help you save a lot of time. Fortunately, phones and many vehicles have built-in GPS. Simply enter a waypoint to get started. A handheld GPS is recommended if you intend to walk a significant distance from your vehicle.

Many guidebooks include GPS coordinates for sites, but not all of them do so in the same format. Make sure the format of your GPS unit matches the format of the book you choose to follow.

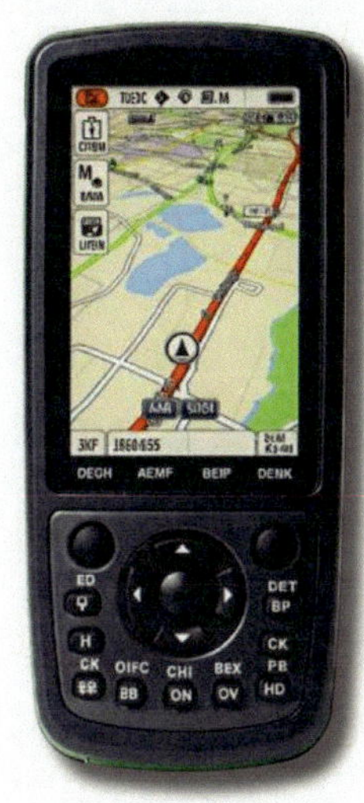

Guidebooks and Maps

No good rockhound is without a plethora of road and geology maps. One of the most crucial rock-hounding skills is the ability to read a geological map. Knowing where minerals can form is a big help when rockhounding.

Bring paper maps with you, as phones and GPS units can run out of battery. Sometimes the road you need to take isn't marked on a map. When you lose cellular signal, your phone's map apps will also stop working correctly.

Guidebooks are also extremely useful, especially when you first start rockhounding. When you first start, it can be challenging to know where to go. Guidebooks contain all the information you need for a successful treasure hunt, including maps, GPS coordinates, and other details. Many guidebooks have been written about rockhounding locations in various states and provinces.

Camera

Rockhounding can take you to some of the most beautiful places you've ever seen—and you'll want to remember them. There's so much more to rockhounding than just the rocks, from breathtaking landscapes to flora and fauna. Bring a camera to capture the beautiful places and sights you'll see. Fortunately, most of us have a camera built into our phones, making it much easier to capture those special moments. If you intend to bring a traditional film or digital camera into the field, consider the environment in which you will be rockhounding. A camera can be damaged if exposed to excessive rain or water drops. Sand can also become trapped within.

Tools and Equipment for Identification

With practice and time, you can identify many rocks and minerals simply by sight. Even the most experienced rockhound can get stumped, so having a well-stocked arsenal of books and identification tools is beneficial whether you're new to rockhounding or have been doing it for years. The following section can teach more about the best identification tools and equipment.

Identification Books and Field Guides

There are nearly as many rock and mineral guides as there are minerals and rocks. Some books cover rocks and/or minerals worldwide, while others concentrate on a specific state, province, or even a single location. Most guides contain the same basic scientific facts, but each includes some information that the others do not. Choose a guide that has good color photos to help with identification. Keep in mind that it is nearly impossible to include a photo of every mineral representation in a single publication. A book containing every type of quartz formation would be enormous. Begin with a general guide and gradually add more specific books.

Magnifying Glasses and Loupes

Magnification is sometimes required to see the finer details of a mineral that aid identification. They enable you to see trace minerals, grain surfaces, tiny crystals, crystal structure details, and microfractures.

Loupes (also known as hand lenses) are small magnification tools that fit your pocket. They are available in 10x, 15x, and 20x magnifications. The lowest magnification, 10x, provides the best depth of field and light capture, whereas 20x is better for viewing fine details.

Ultraviolet Light/Lamp

When exposed to shortwave UV light, longwave UV light, or both, some rocks, gems, minerals/mineraloids, and fossils fluoresce. Some collectors base their entire collection on fluorescent minerals. Serious collectors frequently construct their UV boxes or entire rooms to house their glowing rocks.

Geology Pick

The geology pick (also known as a geologist's hammer, rock hammer, or rock pick) is a tool that is synonymous with rockhounding. Almost every rock club and many rock shops use a silhouette of a geology pick as a logo, to the point where it's almost cliché. A rock hammer typically has two heads, one on each side. One flat, square head is used to break rocks and hammer in chisels (also called gads). The opposite side is either a chisel or a pick. The chisel head is used to scrape away loose rocks, dirt, and vegetation. It can also be used to split slate and shale when looking for fossils. The pick head is used for prying, digging, and splitting rocks. The majority of geology picks are about 13" long.

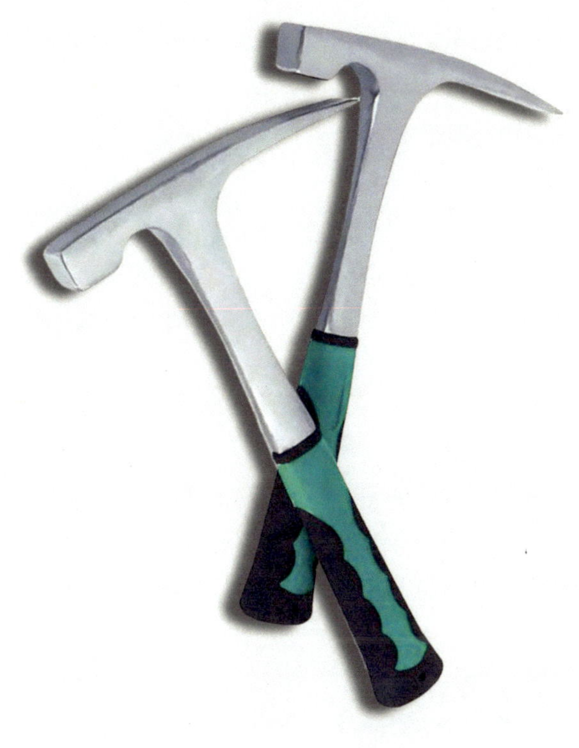

Geology picks are distinct from standard hammers because the steel is specially tempered to pound on a rock. Geology picks are also solid steel rather than a steel head attached to a different material handle.

When used on rocks, regular hammers can splinter and scatter shrapnel everywhere. The hammerhead is typically attached to a wooden or fiberglass handle, which can break when used on rocks, sending the hammerhead flying. When rockhounding, never use a regular hammer.

Don't hesitate to spend money on a high-quality tool. You can have that geology pick for life with the proper care. To prevent rusting, coat your tool with water-resistant oil and sharpen it regularly.

Your objectives determine the type of tool required to collect samples. For example, when searching for small gems, the tools needed may differ from those used today compared to breaking a large rock to extract the minerals it contains. A rock hammer, pickaxe, chisel, and stick are recommended for rock hunting.

Here are the tools that most beginners use, almost in order of importance:

Hand Tools

A variety of small tools are available for precision work. A sieve, a small pickaxe, a spoon, or a small knife can all be used. A rock hammer or a pickaxe is the essential gathering tool every Rock Hound should have. It is suitable for grinding or fine grinding but not for crushing large stones.

Cleaning Supplies

To clean the area while looking for rocks, bring a small broom, brush, and a syringe filled with water. Cleaning the sample allows you to identify it more easily.

Crushing Hammer

A crushing hammer is a miniature version of a hammer. It usually weighs between 2-4 pounds. Most people should be fine with the £3 version, but ensure you finish it quickly. A breaker hammer is frequently used in conjunction with a chisel.

Shovel

At rockhounding sites, shovels are used to dig holes. They are available with both long and short handles. Long-handled shovels are used to dig larger holes in more open areas. Short-handled shovels are ideal for working in small spaces or when you don't need to dig a large hole. Most rockhounding shovels have a spade head, but flat-head shovels can help scrape up the last bits of pay dirt, or gem-filled concentrate, from the ground. When it comes to shoveling handles, fiberglass outperforms wood.

Bring a shovel if you want to move some material up for sampling. Small folding knives are less bulky and easier to transport but more difficult to use. A regular full-size bucket is required for large tasks.

Crack Hammer

Geology picks are great for breaking rock, but if you need to smash up large or tough rocks, you'll need to bring in a crack hammer, a heavy hammer designed for breaking rock and chisel work. The majority of rockhounds prefer to use hand sledges. These are crack hammers with shorter handles that come in various sizes and weights. Most weigh 2-4 pounds and measure 10"-12" in length.

Chisels

Chisels, also known as gads, are pounded into rock with a rock hammer to break them apart. They can be used to shape specimens or split larger rocks or geodes. Chisels are available in a variety of sizes and tip shapes. Chisels with a flat, straight edge are used to break rocks along a line. These chisels are frequently used for geodes and specimen trimming. They can also be used to aid in the discovery of natural fractures in large rocks. Chisels with a pointed tip direct the force of the hammer to a single point.

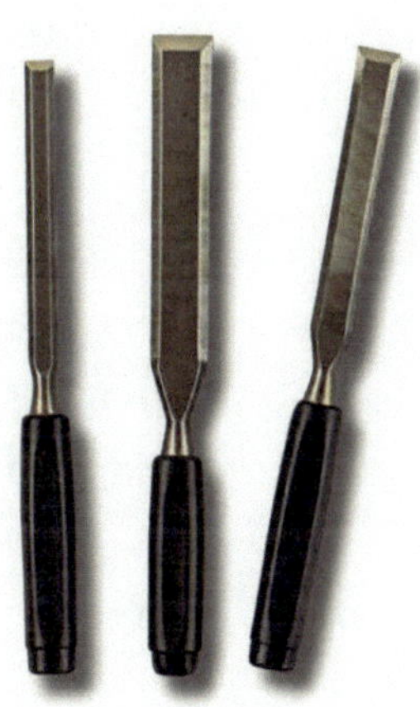

Most people want various chisel sizes, some with wide blades and some with sharp blades. Ensure your chisel is marked for masonry use; wood or metal chisels are insufficient. Chisels with handguards are strongly advised for children to avoid painful hand injuries.

This hammer can crush boulders and is not required for rock movement. It can be difficult to swing and inspect the device comfortably.

When it comes to purchasing chisels, there are two schools of thought. You can buy cheaper chisels if you anticipate losing or breaking many of them, or you can spend more money on higher-quality chisels that will last much longer (but sting a bit more when lost). In the end, it's all up to you.

Large rocks are dislodged and moved using pry bars and crowbars. Their lengths range from 18" to several feet.

The Estwing Gad Pry Bar is an excellent tool for rockhounds. It's an 18-inch-long steel piece with a pointed tip on one end and a flat chisel head on the other. It is handy for hard-rock mining because it is a multi-tool in one.

It's a good idea to have a few different sizes of pry bars on hand for other jobs.

Screens

Sifting screens are used to separate dirt from rocks and categorize rock sizes. They are constructed of galvanized hardware cloth. The most common hardware cloth sizes have 1/4" and 1/2" mesh openings, but larger and smaller screens can be used depending on your needs. For small garnets, I have a small 12" square screen with a 1/2" mesh that I use.

Most screen frames are made of wood, but some are made of PVC piping or aluminum. Smaller screens can be used on the ground with ease. Consider purchasing or building a stand to rest the screen on for larger screens, so you don't have to bend over to shake it. Tripods with a screen attached work nicely as well.

Gold panning involves the use of stackable plastic screens. They enable the classification and separation of larger rocks and nuggets from fine gold.

Screens are simple to make at home or to purchase online. They are also available in some rock shops.

Gem Scoop

A gem scoop is a tool with a long handle and a slotted scoop at the end. It's used to extract mineral specimens from gravel and dirt. Gem scoops come in handy when exploring gravel deposits along oceans, lakes, creeks, and rivers. When it comes to avoiding back strain, gem scoops are a lifesaver. Long hours of bending over to pick up rocks are bad for your lower spine, but gem scoops allow you to stand up straight or slightly bend.

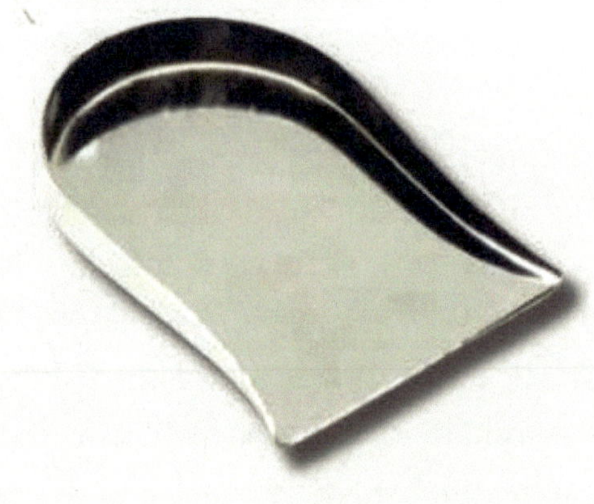

Most commercial gem scoops are made of aluminum, so they are light and can even be used as walking sticks. Because the scoop is not welded to the handle, it should not be used to pry heavy rocks. If you notice your scoop has become loose over a long period of use, you can pinch the metal again where the scoop is attached to the handle.

Probes

A probe is a long piece of metal that has been bent at one end to form a handle. Agate and jasper hunters use them to probe the earth for minerals. This is because agate and jasper emit a distinct "ping" sound when struck by metal. You've found an excellent place to dig if you hear this sound.

Buckets

Five-gallon plastic buckets are a popular tool for collecting and storing rocks. Buckets are useful because they have handles and are easy to obtain, and a full 5-gallon bucket is about the maximum amount of material that most rockhounds can carry. If you can't pick up a full 5-gallon bucket, try a 1- or 2-gallon bucket.

Use a Broll for easier bucket transportation on flat surfaces. It's a wheeled device with a long handle that you attach to a 5-gallon bucket. Carrying a bucket full of rocks is much more complex than rolling the bucket.

If you intend to use buckets for outdoor storage, drill some small holes in the bottom of the buckets to allow water to escape. If a bucket of rocks is not allowed to drain during a rainstorm, it will become coated in slime and algae.

Bags and Packs

Heavy-duty backpacks and bags are other popular items for moving rocks in the field. Make sure the bag you choose is made of a sturdy material that can withstand the weight and sharp edges of rocks without tearing. A metal-framed backpack distributes the weight of rocks on the back.

A cheaper option for lugging rocks around is sandbags. You can get them in bundles of one hundred, they're firm, and when empty, they take up much less room than buckets.

Spray bottles are used to clean dirt off specimens in the field. When rocks are wet, you can see their appearance when polished. Spray bottles are also beneficial to humans. A cool mist of water on your face in a hot desert can be one of the most refreshing sensations you've ever experienced. (Make sure you have plenty of drinking water!)

Brushes

Brushes are another item that can be used to clean dirt off specimens while out in the field. They are available in a variety of sizes and shapes, with a variety of handles. Plastic, steel, aluminum, or brass bristles are all options. Metal bristles can damage some softer minerals, so use a scrubbing stick with a plastic bristle brush if you're unsure about the hardness of the mineral you're cleaning. Pro tip: Repurpose old household cleaning brushes and toothbrushes for scrubbing rocks. Paintbrushes can sweep dirt out of a dig hole or vug.

Tools to Mine Gold/Silver/Platinum

Many books have been written about the tools used in precious-metal mining. A gold pan, sluice box, or metal detector are good places to start for a beginner gold hunter. If you get the "gold bug," you can start looking into more advanced tools and machinery.

Gold pans are the traditional method of recovering gold from rivers and streams. Gold is weighty; by shaking and swirling material from a river or creek, any gold should settle to the bottom while lighter rocks float out of the pan. When you're finished, you should have heavy materials like black magnetite sand, garnets, and hopefully precious metals. The gold must then be separated from the rest of the heavy material.

A sluice box, a long, narrow box with bumps or riffles along the bottom, is another method for separating gold from the dirt. Water passes gold-bearing material through one end of the box and pushes all lighter material out the other.

Metal detectors can be used to find larger gold nuggets and are especially useful when prospecting for gold in arid areas where washing dirt with water is not an option.

You're well on your way to finding buried treasure with a better understanding of rockhounding in your back pocket and the tools for the job in your own rockhound's toolbox! Following that, you'll learn where and what to look for when rockhounding in a forest, on a mountain, or along a river or ocean.

Before you proceed, please keep the following in mind: In addition to the information in this chapter, read any included instructions and cautions that come with the equipment and tools you intend to use. Remember that safety is essential in any enjoyable rock-hounding adventure.

Prospecting tools include:

- A small **shovel**, preferably a military-style folding shovel. Be wary of cheap "camp" shovels made in China. The genuine article will have "US" stamped on it, be heavier, and last much longer.

- A **potato rake** with a shortened handle. Dig into rocky or root-infested soil with this.

- A small, inexpensive camp saw for cutting roots. Don't spend too much money because it will dull quickly and must be resharpened or replaced frequently.

- A 2 or 3-pound one-handed **sledge hammer** for small cracks. They are used to break small rocks or trim large specimens.

- A small **wooden-framed screen**. Use a 12" or 14" "hardware" screen to separate dirt from small rocks (and hopefully gems or desirable materials).

- **Small pick**. Estwing makes a great one, and you can sometimes find one used for gardening.

This basic prospecting tool kit and essential hiking equipment should fit into a small, lightweight, and comfortable backpack.

- At least 1 liter of drinking water
- Repellent for insects
- Work gloves made of leather
- Standard first-aid kit
- Standard survival kit (knife, fire starter, compass, whistle, disposable rain poncho)

Always notify someone of your plans. Carry a GPS and a cell phone, but don't rely on either. Stay in one spot if you get lost.

Moose are common in Maine's woods. Enjoy them from a safe distance. Seeing black bears, on the other hand, is extremely rare. It is even more unusual for a bear not to flee at the sight of a human.

In May and early June, biting insects such as mosquitoes, black flies, deerflies, and no-see-ums are ferocious; these flying annoyances can be unbelievable. Always bring insect repellent — the real stuff contains "DEET." Keep an eye out for bees and hornets and a safe distance. Ticks can be found anytime there is no snow on the ground, and Lyme disease is a problem in Maine, so prepare yourself.

Remember that poison ivy, oak, and sumac can be found almost anywhere.

Following these will help keep you out of trouble. Accidents happen, and nothing is guaranteed. The author accepts no responsibility for mishaps in the Maine wilderness. Wild animals are inherently unpredictable, and other potential threats are too numerous to list.

Equipment for Safety

Your top priority should be safety. Because you are hunting for rocks in remote areas using various tools, rockhounding can be a dangerous pastime. So, research any hazards associated with the area you are visiting and be ready to use special safety equipment if necessary. Here are a few:

- **Work Gloves**: These protect your hands from cuts and scrapes while keeping them clean.
- **Ankle Boots**: Appropriate for mountains and other uneven terrains. Wear waterproof rubber boots if you intend to be in the water.
- **First Aid Kit**: Because you are not near medical care, treating minor injuries promptly and appropriately is critical.
- **Helmet**: It is recommended that you wear a helmet if you intend to hunt in caves, under rocks, or in other places where rocks fall.

Of course, you'll need a way to transport your sample home after a successful rockhounding day. Carrying heavy specimens by hand is best avoided to keep the specimens in pristine condition. It is also for your safety.

While any container can be used, there are a few options you can choose from to help you take your samples home. At the same time, carrying a few simple tools can assist you in determining what you've discovered before taking it home. Here are a few of them:

- **Backpack**: The majority of Rockhounds carry their gear in bags. Choose one that will not scratch easily with a sharp tool and has enough space for your needs.

- **Bucket**: This is a good choice when there are a lot of specimens. Their size and quality vary. Buckets can handle heavy loads. Choose one with a comfortable grip.

- **Loupe**: This tool allows you to see the details of your sample field quickly. 10x power is usually inexpensive and sufficient for identifying rocks.

- **Magnets**: These can be used to identify meteorites and ferrous rocks like hematite and magnetite. Refrigerator magnets or other disposable magnets can be used for this.

- **Field Guide**: A guide used to identify and classify various types of rocks, gems, or minerals.

- **Spray Bottle Filled with Water**: Because samples are frequently dusty, dirty, and grimy, it is hard to see them. Using a spray bottle and some used rags, you can clean them for easier identification.

CONCLUSION

I hope this book was able to help you to gain some insight into the world of Rockhounding. There are many different types of stones in this world, but this book is an excellent place to start learning about their properties and usages across the globe.

When you first start, there's a lot to learn, but it doesn't have to be frightening. You can begin with simple equipment and research. Use all available rockhound resources, including books, clubs, museums, universities, and even shops, to learn all you can about rockhounding. Before embarking on your third or fourth trip to a rock heap, you will become acquainted with the tools for the various collection methods and develop a taste for rock and minerals.

We encourage Discover Nature in the Rocks readers to use their powers of observation. As a result, this book includes activities and experiments that can be done both inside and outside.

When carrying out the experiments, exercise extreme caution. They have been tested, but accidents can happen at any time. Use common sense and a healthy dose of safety consciousness: Wear safety glasses when necessary, gloves to protect your hands, a buddy system when exploring outdoors, and be wary of heat and cold.

This rewarding fun will grow over time, providing you with hours of entertainment and the opportunity to meet new people.

Consider this the first step toward becoming a certified gemologist now that you have a basic understanding of some of these stones and their various varieties. Whatever your reason for wanting to know more about these stones, keep learning! This is just the beginning.

If you intend to use these stones for healing, that's great! Purchase some immediately. Based on the information in this book, you should be able to tell the difference between a fake and a genuine one.

If you are looking for a stone to help you in a specific way, you can make a more informed decision. Many people in this world will give you incorrect information about these stones. Often, it is not because they are malicious but because they are simply misinformed. You do not have to be one of them! Use this book's information to educate and assist others.

Have fun and stay safe!

Made in United States
Troutdale, OR
12/20/2025

44045870R00074